Stetson University
3 4369 00411234 2

D1000345

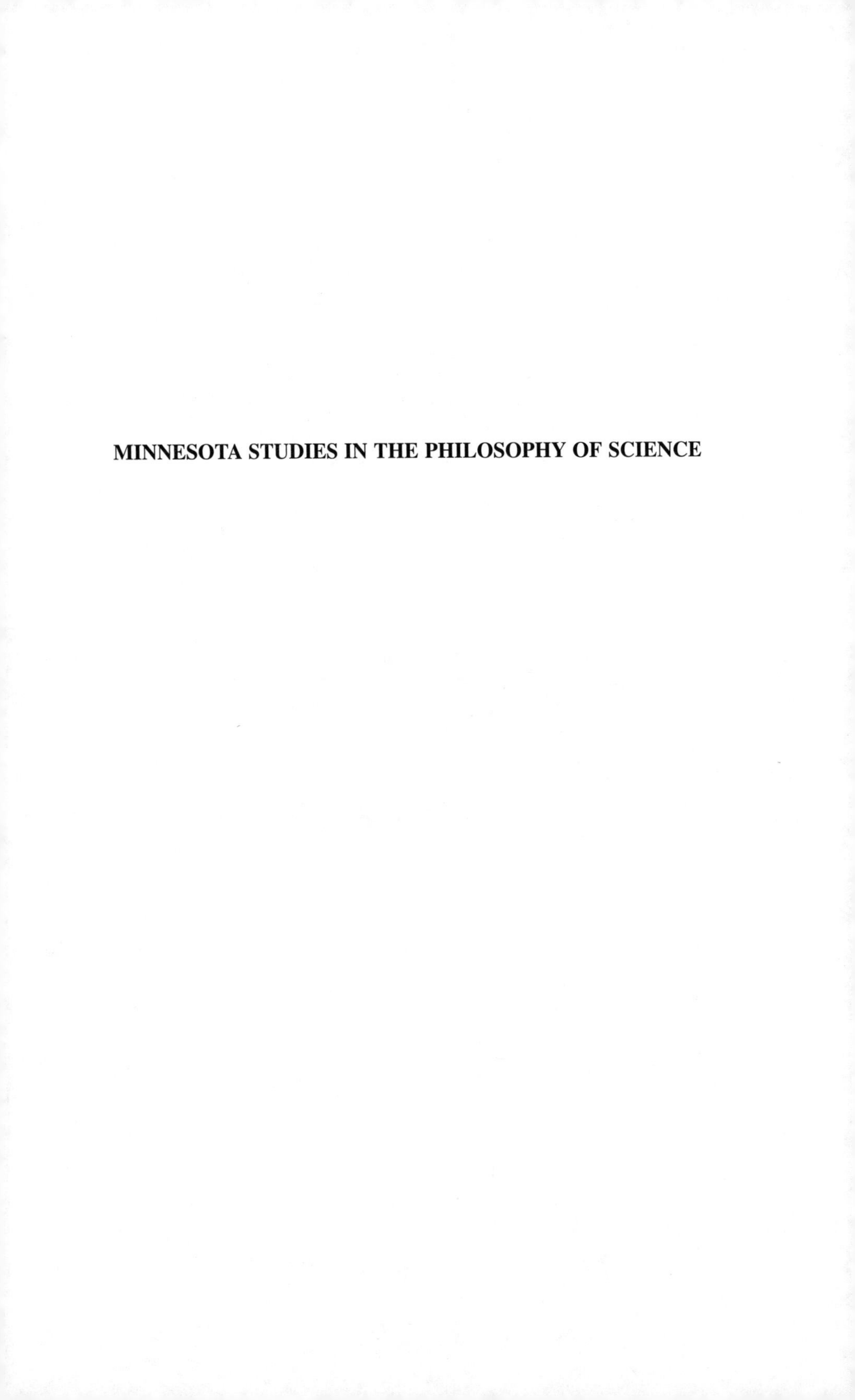

MINNESOTA STUDIES IN THE PHILOSOPHY OF SCIENCE

Minnesota Studies in the PHILOSOPHY OF SCIENCE

C. KENNETH WATERS, GENERAL EDITOR

HERBERT FEIGL, FOUNDING EDITOR

VOLUME XVII

Quantum Measurement: Beyond Paradox

EDITED BY

RICHARD A. HEALEY
AND
GEOFFREY HELLMAN

University of Minnesota Press
Minneapolis
London

Sections of "Varieties of Quantum Measurement," by William G. Unruh, were originally published in *New Techniques and Ideas in Quantum Measurement Theory,* edited by Daniel Greenberger, New York Academy of Science, 1986; reprinted with permission.

Published by the University of Minnesota Press
111 Third Avenue South, Suite 290
Minneapolis, MN 55401-2520
http://www.upress.umn.edu

Printed in the United States of America on acid-free paper

Library of Congress Cataloging-in-Publication Data

Quantum measurement : beyond paradox / edited by Richard A. Healey and Geoffrey Hellman.
p. cm. — (Minnesota studies in the philosophy of science : v. 17)
Includes index.
ISBN 0-8166-3065-8 (hc : alk. paper)
1. Quantum theory—Congresses. 2. Physical measurements—Congresses. I. Healey, Richard. II. Hellman, Geoffrey. III. Series.
QC173.96.Q363 1998
530.12—dc21 98-17341

The University of Minnesota is an equal-opportunity educator and employer.

10 09 08 07 06 05 04 03 02 01 00 99 98 10 9 8 7 6 5 4 3 2 1

Contents

Preface

This volume had its origins in a workshop on quantum measurement held by the Center for Philosophy of Science at the University of Minnesota in May 1995. This workshop brought together leading philosophers of physics and physicists sharing common interests in the long-standing fundamental problems associated with quantum measurement.

In an earlier era of "natural philosophy," physics and philosophy of physics were quite inseparably intertwined, but in the modern age of proliferating specialization, fruitful communication across the disciplines has become the exception rather than the rule. We would like to think that the workshop and this volume are symptomatic of an ongoing process of reunification, one which can pave the way toward exceptional progress in this fundamental and highly challenging area, and others as well.

As the opening chapters of Leggett and Shimony bring out, standard conceptions of quantum mechanics (QM) and its universal applicability prima facie conflict with the definiteness of macroscopically distinct states (of instruments or even ourselves) which we commonly experience and tacitly appeal to in the very enterprise of gathering statistics to test QM itself. One possibility is that QM simply is not really universal, that between micro- and macrosystems "something else"—as yet unknown—happens. Leggett outlines a program to gain an experimental hold on this, revealing a beautiful connection with the famous work of John S. Bell on experiments of the sort contemplated by Einstein, Podolsky, and Rosen (in the mid-thirties) and by Bohm (in the early fifties). Whatever the outcome of the experiments described, we would appear to have a stunning result: either QM would actually be shown to fail at a sufficiently coarse-grained level, calling for qualitatively new physics; or the conflict between QM as ordinarily understood and macroscopic realism would be revealed to be even deeper than we had contemplated.

A famous line of thought going back to Hugh Everett's "Theory of the Universal Wave Function" (in the mid-fifties) begins with universality. Everett and several of his successors have made ingenious efforts to "translate" statements relating our common experience of definiteness of experimental outcomes into the language of wave mechanics, with some apparent, if surprising, success. Bub, Clifton, and Monton subject some recent such efforts to critical scrutiny and show that they can achieve at best partial success, necessarily omitting crucial information concerning the particular outcome events themselves.

Although the outcome of Leggett's proposed experiments should not be prejudged, failure of QM can hardly be expected. All this, then, strongly motivates exploring a new course, that of attempting systematically to supplement the QM description of physical reality with principled assignments of definite properties in circumstances—notably at the conclusion of good measurement interactions—in which QM, as ordinarily understood, withholds them. This involves severing the so-called eigenvalue-eigenstate link of orthodoxy, but raises the challenge of finding a "safe set" of value assignments, free of inconsistencies known to arise whenever too liberal a policy of supplementing QM is followed (the upshot of the corollary to Gleason's theorem, the theorem of Kochen and Specker, and related "no hidden variables" proofs). Healey's chapter presents an overview of this program of "modal interpretations," which forms the central core of this volume. As that essay brings out, a successful resolution of the measurement problem along these lines will likely call upon the work of physicists on decoherence phenomena (the rapid diminution of quantum interference terms due to interactions with environment, touched on in the chapters of Elby and Unruh as well as that of Leggett), which can play an important role in helping select a "safe set" of empirically correct definite properties. As Healey describes, decoherence can help overcome a major challenge to modal interpretations raised by the widespread phenomenon of imperfect measurements. As Elby's contribution brings out, however, the options available a priori for making metaphysical sense of the quantum world are limited, and it is extremely difficult to see how decoherence phenomena all by themselves are up to the task. This reinforces Healey's suggestion that decoherence coupled with a modal interpretation provides a powerful tool at this fundamental level.

Thus, resolving the measurement problem will also call upon considerable theoretical innovation, and if it is to be within a unified quantum mechanical framework as we are now contemplating, it may well be along the lines exhibited by the chapters that follow, by Bacciagaluppi and Hemmo, Vermaas, Dieks, and Dickson. Each of these can be seen as responding to a further challenge to modal interpretations. In the case of Bacciagaluppi and Hemmo, the challenge is to make sense of state preparation on a modal interpretation in which there is no "collapse of the wave-function" but rather certain value-states in addition to quantum statistical states. Vermaas then addresses the challenge of expanding the core property ascriptions of modal interpretations in mutually consistent ways, thereby bringing a measure of order to a variety of extant proposals. The challenge addressed by Dieks is to choose a privileged way of factoring tensor product Hilbert spaces used to represent quantum states of complex systems (e.g., the universe itself) so that the procedure of assigning properties in a modal interpretation is well defined. Finally, Dickson approaches the challenge to provide a dynamics governing the evolution of synchronically possessed values assigned by modal interpretations. Individually and collectively, these contributions reveal the vitality and resourcefulness as well as the challenges of this exciting new line of research.

If we look at measurement epistemologically, as Unruh does in his chapter, we see a rich variety of ways in which one physical system can be used quantum mechanically to gain information about another, the standard schemes highlighted in the measurement problem spanning only a limited range among the possibilities. While this leaves open the question of how consistently and accurately to model the latter, it does call for a broader point of view if we are to achieve a comprehensive understanding of quantum reality and our own part as investigators within it.

We would like to take this opportunity to express our gratitude for support to the National Science Foundation, the Theoretical Physics Institute at the University of Minnesota, the College of Liberal Arts and the Graduate School of the University of Minnesota, and especially to Ron Giere and Steve Lelchuk of the Center for Philosophy of Science.

GEOFFREY HELLMAN
RICHARD A. HEALEY

Anthony J. Leggett

Macroscopic Realism: What Is It, and What Do We Know about It from Experiment?

1. What Is It?

Perhaps the easiest way to see the motivation for, and the meaning of, the concept I call "macroscopic realism" is to start by discussing what it is not. Of course we are all realists in the context of our everyday lives—how could we not be? Chairs are on this side of the table or the other, meters read 5 amps or 10 amps or some other definite number, particle counters either click or do not, and so on. The difficulty is, of course, that if we take quantum mechanics to be a complete theory of the physical world, then most interpretations (though not all) of the formalism appear prima facie incompatible with this "common-sense" point of view.

Consider a simple generic Young's slits interference experiment done with a single microscopic particle at a time. The system starts from state A and can reach (among other possibilities) state E by either of two paths, passing respectively through intermediate states B and C. If $P_{A \to B \to E}$ represents the probability that the particle, starting from A, arrives at E when the path C is physically shut off (e.g., by obstructing the relevant slit in a diaphragm), $P_{A \to C \to E}$ similarly the probability of arriving at E with path B shut off, and $P_{A \to B \text{ or } C \to E}$ the probability of arriving at E when both paths are left open (note the somewhat tendentious notation!), then under appropriate experimental conditions we may obtain the result

$$P_{A \to B \text{ or } C \to E} \neq P_{A \to B \to E} + P_{A \to C \to E}.$$

Indeed, in some particularly spectacular cases we find that the left-hand side is zero even though both terms on the right are nonzero and positive. This statement about probabilities is, of course, a statement about the properties of the ensemble of identically prepared particles as a whole. Yet we know that if, for an identical ensemble, we place detectors which discriminate which of the (mutually exclusive) intermediate states B or C was chosen, we indeed find that each individual particle realized one of these states or the other. Note that these remarks are purely a description of the experimentally obtained results and do not refer in any way to the standard account given by the quantum-mechanical formalism of the experiments in question, although needless to say they are completely compatible with the latter.

Let's now push ahead naively and ask (referring to the case where no detection of the intermediate state is made), Is it true that, on each individual trial, one or the other of the two possibilities B or C was realized? There are many possible reactions to this question, including the standard Copenhagen one that the question itself is meaningless and should not be asked. Purely for pedagogical convenience let me assume that we decide to regard the question as meaningful and answer it in the negative (I emphasize that nothing in the bulk of this chapter hangs on the assumption, since this part of the argument is of purely motivational nature). Referring to the standard QM description of such an experiment, in which the state description of the intermediate state of the (systems of the) ensemble is a linear superposition, $a\psi_B + b\psi_C$, of the QM probability amplitudes ψ_B, ψ_C for B and C respectively, we would then naturally say:

> When a microscopic system belongs to an ensemble where the correct QM description is by a linear superposition of probability amplitudes for two different states B and C, then it is not true that one or other of these two states has been realized. (Q1)

(I deliberately refrain from trying to cast the statement in a more positive form, such as "both possibilities are still represented.")

It is essential to my (motivational) argument at this point to distinguish carefully between the statement itself, which may be regarded as a proposition about the *meaning* (or lack of it) of the QM description, and the *evidence* for the statement, which consists of (a) the experimentally observed phenomenon of interference and (b) the remark that a standard application of the QM formalism, including the usual measurement axioms, does in fact correctly predict this and other features of the experiment. Once one has been through the relevant arguments for a few prototypical situations like those above, it seems natural to take the above statements as generically true.

Now comes the crunch: With a few provisos which are in essence irrelevant to the present argument, I believe it would be universally agreed that it is possible to construct experimental setups such that, *if* we believe that the formalism of QM is universally valid, it will give a description of the form $a\psi_1 + b\psi_2$, where ψ_1 and ψ_2 are *macroscopically distinct* states of the relevant part of the universe (*including* all relevant parts of the "environment"). Let's call such setups "Schrödinger's-cat" type situations: in real life the states ψ_1 and ψ_2 will most likely not correspond to the living and dead states of a cat but rather, for example, to states in which a given counter has or has not fired, suitably entangled with the radiation field, the air molecules in the surrounding laboratory, and so on. Given this, are we to reaffirm statement (Q1), with "microscopic" replaced by "macroscopic"? ("Macroscopic system" is here to be understood as the whole of the relevant part of the universe.)

With the exception of (some versions of) the "relative-state" interpretation, almost all the currently marketed interpretations of the QM formalism answer this question in the negative; that is, they assert that actualization of one or other of the two macroscopically distinct possibilities has indeed taken place. Justification for this point of view is usually sought in the phenomenon of "decoherence" (Zurek 1991), of which we shall no doubt hear much more; in the present context, this would effectively imply that since the states ψ_1 and ψ_2 are entangled to an extreme degree, no conceivable experiment could possibly exhibit the effects of interference between them. While this assertion is no doubt true, at least in all but very specially engineered cases (see below), its use to deny the macroscopic version of (Q1) seems to me to involve a major logical fallacy. (Q1) is, as emphasized above, a statement about the meaning of the QM formalism. This formalism is itself a seamless whole, extending (at least in the view of most of its practitioners) all the way from the realm of subatomic particles to the macroscopic, everyday world, and therefore any interpretation of its meaning which changes radically between the microscopic and macroscopic levels must violate a principle of continuity. The fact that the evidence for this interpretation, abundant at the microscopic level, has for all practical purposes vanished by the time we get to the level of everyday life, is as a matter of logic totally irrelevant to this consideration.

One way out of this dilemma is to refuse to interpret the QM formalism in any way at all, that is, to deny that the assertion (Q1) has any meaning either at the microscopic or even at the macroscopic level. This is essentially the point of view taken by adherents of the full-blooded "statistical" interpretation of QM, which may be regarded as the logical development of the Copenhagen approach. From such a standpoint, the whole formalism of QM is nothing more than a calculational recipe, which has no meaning at all in its own right: the only point at which contact with physical reality is made is at the very end of the calculation, where predictions are made for the probabilities of *directly observed* macroscopic events. I personally find this (non)interpretation internally consistent but extremely uncongenial, in part because it seems to imply that at least in some circumstances questions about the state of the world, even at the macroscopic level, when it is not directly observed are either meaningless or in principle unanswerable, and my distaste is in no way diminished by the arguments given by adherents of the "consistent-histories" approach (Griffiths 1984) to the effect that these circumstances are few and far between.

The conclusion implicit in these remarks is that if we assume that the formalism of QM applies in unrestricted fashion not only at the microscopic but also at the macroscopic level, we have a prima facie problem in reconciling this assertion with our common-sense, realistic conceptualization of the everyday world. I personally believe that this problem is irresolvable, but it does not matter whether anyone agrees with me. What matters is that it provides the motivation for an alternative conjecture: namely, that QM as presently conceived is *not* the whole truth about the world, but that at some level of physical scale, complexity, or degree of organization there come into play new and currently unknown physical laws whose nature is such as to guarantee that it is never necessary to describe the state of the world, even formally, in terms of a quantum superposition of macroscopically distinct states. Rather, a definite macroscopic state is always realized, in a way that is completely independent of our interpretation of the QM formalism. This is the conjecture that I call "macroscopic realism." It is a very generic negative statement; it essentially asserts that QM superpositions of macroscopically distinct states do *not* occur, but makes no particular commitment as to what replaces them (e.g., as to whether, within a particular macroscopic "branch," the QM description is still good). Also, the conjecture is of little interest—at least to the practicing physicist!—unless we can derive from it experimental predictions different from those that follow from the hypothesis that the standard QM formalism is complete and of universal applicability.

2. Could QM Go Wrong?

The point of view I am proposing—namely, that QM may not be the whole truth about the physical world—is likely to be strongly antithetical to the views of many, so let me address the question of its a priori plausibility or lack thereof.

It is difficult to exaggerate the degree to which reductionism is entrenched in the thinking not only of the twentieth-century physical scientist, but also to a large extent of the twentieth-century man in the street. Here I use the word "reductionism" in a loose and philosophically inexact sense, to mean the belief that the behavior of large and complicated things can in principle be explained entirely in terms of that of the smaller and simpler elements of which they are composed. Forget, for the purposes of the present discussion, all the arguments about whether there are new "emergent" principles which manifest themselves only at a certain level of complexity and cannot be meaningfully analyzed in terms of concepts applicable to lower levels; forget, also, any considerations that may be peculiar to biologically organized and/or conscious systems. How many practicing physical scientists have ever even bothered to ask themselves the question whether an exact solution of the Schrödinger equation—which we all agree gives a brilliant description of the behavior of single elementary particles, atoms, and molecules—for (say) the 10^{23} molecules composing an ordinary liquid would predict the correct behavior in all respects? It is, almost, a premise of our professional careers (at least for us condensed-matter theorists) that this is so, and to question it seems comparable to questioning the principle of induction.

Could this proposition be false? I will postpone to the next section the question of experimental evidence one way or the other, and address here the issue of our a priori expectations. One way in which the proposition might fail is that there might be corrections to the Schrödinger equation whose effects are unobservably small at the atomic level but become dominant by the (reasonably) macroscopic level. As is well known, a theory of "actualization" with this character has been constructed by Ghirardi, Rimini, Weber, and Pearle (GRWP) (Pearle 1989), and appears to be internally consistent at least at the nonrelativistic level. Some particle physicists (e.g., Ellis et al. 1984) have also speculated that there may be non–quantum-mechanical effects which originate at the Planck scale, are present but impossible (or extremely difficult) to observe at the scales probed by current accelerators, and have possibly important consequences by the macroscopic scale. Any such hypothesis would not, of course, offend against the letter of the general reductionist prejudice, although one could perhaps argue that it is contrary to its spirit, particularly if it should turn out that the atomic-level effects are so tiny as to be unobservable in any foreseeable future. A more intriguing possibility is that reductionism is wrong *tout court*, that it simply is not possible, even in principle, to understand in all respects the behavior of complex, macroscopic bodies in terms of the behavior of the microscopic entities of which they are composed. No doubt such a hypothesis looks bizarre—and indeed there would be little incentive even to consider it in the absence of the quantum measurement paradox—but it is arguable that it is no more bizarre than appeared, prior to 1956, the supposition that Nature could know the difference between right- and left-handed coordinate systems.

One counterargument that is often advanced in this context runs as follows. We know (it is said) that quantum mechanics works from the Planck scale up to the atomic or molecular scale (twenty-five orders of magnitude); what is so special about the extra ten orders of magnitude needed to get up to the "human" scale that it should suddenly fail? If I were to argue that the difference is that in the first case one is essentially dealing with single elementary particles and in the second with large collections of such, it would be (correctly) responded that from the points of view of modern quantum field theory there is no such thing as an "elementary" particle, that the relationship between the nucleon and the quarks and gluons which compose it is in principle no different from that between the nucleus and its constituent protons and neutrons, or between the molecule and its constituent atoms. Moreover, even the concept of the number of constituents is ill-defined at the subnuclear level (consider, for example, the "strange quark sea" now believed to be an important constituent of the nucleon).

In response, a minor and a major point. First, contrary to the apparent belief of many theoretical particle physicists and cosmologists, we *do not* know that QM works "down to the Planck scale." Rather, most of the relevant community *assumes* that it does, and constructs on this basis various speculative cosmologies. Because the comparison with experiment of those predictions that depend essentially on quantum-mechanical considerations is, to put it charitably, ambiguous, one can hardly claim that the assumption has been verified; the best one can say is that so far no clear evidence has emerged against it (and thus, perhaps, that the principle of Occam's razor makes it reasonable to use it as a working hypothesis).[1] The more important point is that one would not expect that by inspecting the quantum field theory paradigm from the inside one could determine ahead of time whether and where it is likely to break down. It is indeed a seamless whole, and once one has decided it is the whole truth of course the regime between 1Å and 1m is no different from any other. So what? Imagine going back to the year 1895 and telling one's colleagues that classical mechanics would break down when the product of energy and time reached a value of order 10^{-34} joule seconds. They would no doubt respond gently but firmly that any such idea must be complete nonsense, since it is totally obvious that the structure of classical mechanics cannot tolerate any such characteristic scale!

3. Where Is the Evidence?

Just about everyone agrees that standard quantum mechanics based on the Schrödinger equation (or, when necessary, on the Dirac equation or some other relativistic generalization) works just fine at the level of single atoms, molecules, and (at least over a reasonable range) subnuclear particles; and we already agreed that in the absence of evidence to the contrary, the principle of Occam's razor indicates that it is sensible to assume, as a working hypothesis, that its validity extends down to the Planck scale. If we put aside for the moment any qualms concerning the quantum measurement paradox, a similar argument would indicate that we should assume that quantum mechanics works, in all respects, right up to the macroscopic scale; needless to say, this is exactly what almost every condensed-matter physicist does in his or her everyday working life. I already discussed, in the previous section, the question of how such an assumption is forced on us a priori; in this section I will examine how far existing experiment provides arguments in its favor that go beyond the simple Occam's razor consideration.

We should first define more carefully the question we wish to answer. The simple question "Does QM continue to work at the macroscopic scale?" is ambiguous in the extreme, a fact that has unfortunately caused a certain amount of avoidable confusion in the literature of the past dozen years. I will use the phrase in one sense only, namely: "Does the QM formalism continue to give correct results even when it predicts superpositions of macroscopically distinct states?" (We will define "macroscopically distinct" below.) Note that prima facie the answer "yes" to this question appears to exclude the hypothesis of macrorealism as introduced in the first section. Strictly speaking, this is not so: it could perhaps be that, at least in a large class of experimental situations, the hypothesis of macrorealism is correct but nevertheless the QM formalism continues to predict correctly the experimental behavior. This would be analogous to the conjecture, conclusively disproved by the late John Bell (Bell 1964), that while the QM predictions for two-photon correlations in the "Einstein-Podolsky-Rosen" (EPR) situation are in agreement with experiment, the true underlying theory is of the local hidden-variable type. At any rate, just as even in the absence of Bell's theorem it was of interest to try to verify the somewhat counterintuitive aspects of the QM predictions for the EPR experiment, so in the present context it is interesting to ask what positive evidence, if any, we can produce for the occurrence of the macroscopic superpositions predicted under appropriate conditions by the quantum formalism; this is independent of the question of whether we can use this evidence to exclude macrorealism.

We need a more precise definition of "macroscopically distinct," and in fact it will be helpful more generally to try to define a quantitative measure of the "distance," as it were, along the axis that takes us from well-verified single-particle interference experiments of the Young's slits type to the paradoxical situations of which Schrödinger's cat is paradigmatic. Various possibilities exist for such a "measure," an obvious one being related to the difference in displacement of the system center of mass (COM) in the two branches of the relevant wave-function; it is a suitably macroscopic value of this quantity that triggers the reduction (actualization) process in the GRWP theory (consider also the scenario proposed by Penrose [1986]). Unfortunately, most of the relevant experiments conducted to date (see below) do not involve any substantial displacement of the system center of mass in either branch, so it is advantageous to generalize the definition. One possibility might be to consider, in addition to COM displacement, other quantities that can take macroscopic values such as the magnetic moment distribution and the mean electric current, among others, and to define the relevant measure as the maximum value obtained within this class of the difference (expressed in suitably dimensionless terms) of the expectation value in question between the two branches. An alternative approach, which I believe is for practical purposes virtually equivalent, is to try to specify quantitatively a measure of the extent to which large numbers of particles (here taken to be nuclei and electrons) are behaving differently in the two branches. Some years ago I gave a possible definition (Leggett 1980), in terms of the various reduced density matrices of the system, of such a measure, christening it the "disconnectivity" of the quantum state in question; there is little point in going into the technical details here, and it is sufficient to remark that the concept appears to be essentially equivalent to the notion of "degree of entanglement" which has recently received discussion in the context of generalizations of Bell's theorem. For example, a Hartree type of wave-function has by definition a disconnectivity (D) equal to 1, a typical EPR-Bell state has D = 2, a GHZ state (see Greenberger, Horne, and Zeilinger 1989) has D = 3, and the putative final state of the unfortunate Schrödinger's cat has D of order 10^{23}. Although I will for definiteness use this choice of measure in the subsequent discussion, I believe it is unlikely that anything important hinges on this.

My first and perhaps most important point is that if we put aside for the moment the "purpose-built" experiments I shall discuss below, nothing in condensed-matter physics provides any evidence for states with a value of D anywhere near what could be reasonably called macroscopic. Indeed, although it is true that quantum-mechanical considerations have provided brilliant explanations of the behavior of a vast variety of condensed-matter systems, close examination shows that, with a very few exceptions, (a) what the experiment actually measures is a sum of one- or (more commonly) two-particle correlation functions, and (b) in the calculation of these functions it is almost unheard of for the theorist to invoke states with D greater than some very small number (generously, 5 or 6). In fact, quite the opposite is the case: a large part of theoretical many-body physics consists in effect of finding ways to justify the neglect of the subtle correlations associated with high D! I know of only one set of papers in the literature

(Chechetkin 1976, 1982) which seriously tries to invoke states with D ~ 10^{23} to explain the properties of a condensed-matter system (actually superfluid ^{3}He), and these papers were promptly and, to my mind, conclusively shown to be in error (Yip 1984).

A brief note on the so-called macroscopic quantum phenomena (superconductivity, superfluidity, lasers, and so on), which still seem, alas, after fifteen years to be capable of causing confusion in this context. Despite the unfortunate tendency in the literature to refer to the order parameter $\psi(\underline{r})$ that characterizes these systems as a "macroscopic wave-function," superpositions of two different $\psi_i(\underline{r})$ [$\psi(\underline{r}) = a\psi_1(\underline{r}) + b\psi_2(\underline{r})$] as found e.g., in the Josephson effect do not correspond to superpositions of macroscopically different states and do not possess large values of D. Rather, they correspond to states in which a macroscopic number of "particles" (for example, ^{4}He atoms, Cooper pairs) occupy a single microscopic state which may itself be written as a superposition. Thus they are in themselves totally irrelevant to the issue under discussion (although, as it happens, the Josephson effect turns out to be a key ingredient in some experiments which *are* relevant [see below]).

Next, let us consider briefly whether the "Bell's theorem" experiments offer any evidence for high-D states. It is fairly clear that they do not; indeed, most discussions of the implications of these experiments implicitly assume that a "measurement" has taken place as soon as the photon in question has entered the cathode of the relevant photomultiplier, and it is clear that such an assumption, while no doubt question-begging from the point of view of quantum measurement theory, is completely consistent with the experimental results. But this assumption is equivalent to the statement that each macroscopic counter is always in a definite macroscopic state, so that there can then be no question of any Schrödinger's-cat-like entanglement. This argument does not show that such entanglement is in fact absent, only that its absence is completely consistent with the experimental data.

It should be clear, then, that states with a macroscopic degree of disconnectivity are, at least, not staring us in the face and that we shall have to work quite hard to find them, if they indeed exist. So let's start at the other end and ask: For how large a value of D do experiments at the atomic or molecular level offer direct or circumstantial evidence? In other words, if we decide in advance to interpret the experimental data according to the standard prescriptions of QM, and count particles in the usual nonrelativistic way, what is the largest degree of entanglement (disconnectivity) that needs to be invoked to explain the data?

Almost certainly, the largest value of D for which we have direct evidence (in the sense that we can show rigorously that a smaller value would be incompatible with the data) is 2, in the EPR-Bell experiments. (Any QM description of the system with D = 1 could be mimicked by a local hidden-variable theory and is thus excluded by Bell's theorem.) If a GHZ-type experiment can be realized (and the outcome agrees with the QM predictions), we will be able to push the value of D up to 3, and conceivably eventually a little higher, but it is obvious that the realization of any reasonably macroscopic value along this direction is impossible in the foreseeable future.

If we are content with circumstantial evidence, the situation is a little brighter. For example, the fission of Pu (generally believed to involve collective motion of a large fraction of the nucleons) could be regarded as circumstantial evidence that QM indeed correctly describes the correlated motion of ~200 nuclear particles. More directly, a one-slit diffraction experiment has been conducted with K atoms, and Na atoms have been successfully diffracted from the "grating" provided by a high-power optical field. If we describe these experiments by standard QM, then the intermediate states certainly involved states with D equal to the total number of particles (~70 for K, if we count the individual nucleons separately).

Finally, some elegant spin-echo type experiments in chemical physics, if interpreted in the standard way, are evidence for values of D in the nuclear spin system which are comparable to those just discussed (see Pines 1988).

We see that the evidence accumulated so far that "quantum mechanics continues to work at the macroscopic scale" in the sense defined earlier is essentially nonexistent: 200 is a very, very long way from 10^{23}, and any theory of the GRWP type is likely to have little difficulty finding adequate room for a qualitative change somewhere along the way. Moreover, of the experiments cited, only those of the EPR-Bell type actually exclude a "realistic" account. It would be possible to pursue these avenues further (for example, diffraction of light molecules/biomolecules/small viruses) but given current experimental capabilities progress is likely to be unspectacular. (But as suggested by Shimony in the next chapter of this book there may be other good reasons for pursuing the "biomolecular" avenue.)

A further possibility that has been suggested recently is that one might be able to produce states with a high degree of entanglement (high-D states, in my language) by using micromaser cavities. No such experiments have to my knowledge yet been performed, and I suspect that in practice the value of D so obtained is likely to be only marginally greater, if at all, than in the experiments already cited, so that talk of "optical realizations of Schrödinger's cat" or similar seems to me to be stretching language.[2]

4. QM of Macrovariables

Rather remarkably, a different and in some sense much more ambitious approach seems to be more promising. The most obvious way of obtaining states with a macroscopic value of D is to consider a system which possesses one or more macroscopic variables and to try to engineer things in such a way that this variable behaves in a characteristically quantum-mechanical way. We would look for evidence for linear superpositions of states corresponding to "appreciably" different values of this variable and hence macroscopically distinct states. Practical examples of such a variable include the magnetic flux trapped through a so-called rf SQUID (superconductivity quantum interference device) and the magnetization of a single magnetic grain.

Let's discuss a few obvious a priori objections to this program. First, a rather generic objection: any such attempt is bound to employ a condensed-matter system (and probably a rather messy one at that), where it is virtually unthinkable that we can write down a microscopic Hamiltonian in the way in which we are accustomed to doing it, with some confidence, for systems at the atomic or subatomic level. To those trained only in molecular, atomic, or particle physics this situation is one of unspeakable horror, and many find it difficult to believe that results of fundamental significance could ever emerge from such apparently ill-characterized systems. In fact, the situation is not half as bad as it looks: most of what we don't know doesn't hurt us, and in particular if we reliably know the classical dynamics from experiment we can go a long way toward making confident QM predictions.

A second objection is more concrete. Didn't Niels Bohr show us that the motion of a macroscopic variable must always be so far into the quasi-classical limit that the predictions of QM simply reduce to those of classical mechanics (in which case they would clearly be of no interest in the present context)? No. What Bohr actually argued (not rigorously, but let us grant the conclusion for the purposes of the present argument) was that *when the action S is large in unit of h*, then the predictions of QM and of classical mechanics (usually) coincide. The missing step in the argument is that the motion of any variable that is by a reasonable criterion macroscopic must involve an action which is many times h. We can make this true by definition, but this seems a pointless exercise; if we do not do so, then there seems to be no good reason to deny the status of macroscopic variable to, for example, the magnetic flux trapped through an rf SQUID, and it is then easy to show that the motion of such a quantity need *not* involve an S which is many times h.

Third, it may be objected that the mere fact that two states have a macroscopic degree of entanglement makes it hopeless and pointless to look for effects of their superposition, because to see such effects we would have to measure correlations of order of the relevant D, which is in practice out of the question. This would be true if all relevant experiments were of the Bell-GHZ type; however, it is easily seen that this is not so, even at the atomic level. For example, diffraction experiments can reveal quite substantial ($\sim$70) values of D, but we never measure a 70-particle correlation function directly! Although we experimenters have no operator at our disposal that measures N-particle correlations, Nature does: the operator $\hat{U} \equiv \exp -i\hat{H}\,t$! It is easily seen that in a diffraction experiment with Na atoms, for example, it is precisely this feature which is giving us our effect, and it is clear that this principle can be generalized to arbitrarily "macroscopic" systems (Leggett 1980).

The most serious objection to our program is connected with the phenomenon of decoherence and related problems associated with uncertainty of initial conditions. Indeed, the literature of the past fifty years is full of papers that claim to demonstrate, in varying degrees of generality, that these effects will automatically destroy the possibility of any observation of quantum interference between macroscopically distinct states (hereafter abbreviated QIMDS). Why are these papers all wrong—or at least misleading in the claimed generality of their conclusions? To discover why, it is essential to go beyond the hand-waving, back-of-the-envelope arguments given in most of the literature and do an actual calculation for some reasonably plausible model of a real physical system. When one does so, one discovers a number of features that have not been widely appreciated in the literature: (1) The fact that macroscopic bodies have incredibly closely spaced energy levels—sometimes cited as killing any possibility of QIMDS—is largely irrelevant, since most of them are effectively decoupled from the variable of interest; (2) Uncertainty about the microscopic initial conditions is similarly much less important than it has often been taken to be, since, again, it usually does not affect the motion of the relevant macroscopic variable appreciably; and (3) Most important, even quite strong coupling to the environment does not automatically lead to the destruction of QIMDS.

For a typical case of a macroscopic variable coupled to its environment, much of the effect of the coupling is adiabatic in nature and does not result in the permanent destruction of interference; such destruction can result only from nonadiabatic, that is, dissipative processes, and these may be extremely weak and (more important) are calculable. We should note an important ambiguity in the concept of decoherence, which although it is gaining appreciation among the cognoscenti does not yet seem to have filtered through to the general public (Leggett 1990; Unruh 1995). If we consider the reduced density matrix $\hat{\rho}$ of a system coupled to a complex environment, it is tempting to define decoherence as having taken place as soon as the off-diagonal elements of $\hat{\rho}$ have fallen below some specified (small) value. Yet if this is the definition it emphatically does *not* follow that decoherence, however complete, guarantees the nonexistence of QIMDS at any later time. As an example, consider a simple two-state system with tunnel splitting Δ, coupled in the standard way to an oscillator bath with a lower frequency cutoff $\omega_{\min}$ such that $\omega_{\min} \gg \Delta$. It is easy to solve for the dynamics of such a system and to convince oneself that for much of the time the off-diagonal elements are suppressed by a Franck-Condon factor that may be extremely small. However, the effect is not to suppress the two-state oscillations but only to lengthen their period (by the inverse of this small factor). It turns out that this situation is quite generic in the quantum motion of a macroscopic variable (and even that in many cases, for reasons which are too technical to mention here, the period is not always that much lengthened).

Thus, only the dissipative effects of the coupling to the environments are deleterious to QIMDS. This would still not be too happy a situation if we had to try to calculate these effects a priori from some general microscopic Hamiltonian; for, if QIMDS should not be observed where it is expected, a defender of QM need only argue that we had got the Hamiltonian wrong and thus badly underestimated the dissipation. What really saves the program is the possibility of making reliable QM predictions not on the basis of a microscopic Hamiltonian but purely on the basis of a knowledge of the classical motion of the system. This is again a rather technical subject (Caldeira and Leggett 1983, appendix C; Leggett 1984), and I will merely say that I believe there are extremely strong arguments that, given a few generic and very plausible assumptions about the general nature of the microscopic Hamiltonian (but emphatically *not* about its detailed form), one can always do this. Thus, a failure of the experimental results to conform to the predictions of a quantum-mechanical calculation cannot be dismissed as due to the "messiness" of the experimental system and/or our consequent lack of knowledge of its Hamiltonian.

I now turn to the existing experimental results in this area. The workhorse systems for such experiments are superconducting devices that exploit the Josephson effect: in the conceptually simplest experiments, the macroscopic variable is the magnetic flux trapped through an rf SQUID ring (or equivalently the current circulating in the ring), although some of the most conclusive experiments have actually used a related but conceptually rather trickier system, a single current biased Josephson junction. Most of the experiments have measured the rate of escape of the relevant macroscopic variable from a metastable state into the continuum by quantum tunneling through a barrier (known in the trade as macroscopic quantum tunneling, or MQT); the dependence of the rate on parameters such as barrier height is precisely what is predicted by a QM calculation and rules out, in the opinion of most people (including me), the possibility that the effect seen is a spurious one due to nonequilibrium electronic noise or similar factors. What is particularly striking and satisfying is that by using the technique described above to relate the classical and quantum motion, one can predict the tunneling rate, even for quite strongly dissipative systems, *with no fitted parameters*; and that, modulo some relatively small discrepancies in the WKB prefactors and others in some experiments that are not currently understood, the experimental results agree very well with the theoretical predictions (Clarke et al. 1988). This would suggest rather strongly (a) that QM is still working at the level explored in these experiments, and (b) that our current understanding of how to incorporate the effects of a dissipative environment is essentially correct, not only qualitatively but quantitatively. This latter point is of special importance in view of the history of arguments about the destructive effects of decoherence.

It should be added that while the straight MQT experiments are the most impressive as regards the quantitative fit between experiment and theory, a number of other types of essentially quantum behavior of the flux have been observed in Josephson devices, including resonant activation and relaxational tunneling in a two-well system. A particularly pretty demonstration of the effect of level crossing in a two-well system has recently been given by Rouse et al. (1995). However, while the qualitative behavior in these cases is exactly as predicted, it is usually not possible to attempt a parameter-free fit because of a lack of knowledge of some of the relevant classical parameters.

An alternative system that has been explored in the past few years is magnetic grains or biomolecules (Gunther and Barbara 1995). In this case the macroscopic (mesoscopic?) variable of interest is the collective magnetization. In contrast to the Josephson device experiments, which are invariably done on a *single* physical system using in effect a "time ensemble," the magnetic experiments are typically done on a large ensemble of grains, and only the *average* magnetization is measured (only very recently has it become feasible to make measurements on single grains). This feature makes this kind of system unsuitable for crucial tests of macroscopic realism. Many of the MQT-type experiments done on Josephson devices have been repeated using magnetic materials; while there is qualitative agreement with the predictions of QM, these systems are relatively much worse characterized and one cannot attempt a parameter-free fit.

One phenomenon that has not (at least in my opinion and that, I believe, of the majority of the community) so far been observed in the Josephson-device area is the two-state oscillation, which is the macroscopic analog of the NH_3 inversion resonance (macroscopic quantum coherence or MQC). Exactly such a phenomenon is claimed by Awschalom et al. (1992) to be observed in their resonance and noise experiments on the magnetic biomolecule horse spleen ferritin; if this is correct, it would correspond to a motion which involves superposition of states different in the behavior of around 5000 atomic spins (the number of Fe atoms in the ferritin case). However, the interpretation of the raw data given by Awschalom et al. has been severely criticized (Garg 1993), and in view of the relatively ill-characterized nature of the experimental system it will probably be some time before the issue is resolved.

What conclusions can be drawn from these experiments? First, everything seems consistent with the hypothesis that QM is working just as well at this level as on the atomic scale. As to the value of D (disconnectivity) implied, in the case of the ferritin experiments this is clear-cut: if the interpretation of the phenomenon seen as MQC is correct, it certainly corresponds to a value of D of order 5000 (and whether that is called "mesoscopic" or "macroscopic" is a matter of language, not of physics!). The case of the Josephson experiments is a bit trickier, because of the possible ambiguities in the definition of D when gauge fields are involved, but by any reasonable definition it is at least 10^{12} and probably larger. (The difference in the values of flux between which tunneling takes place is of order 0.05–0.2 times the flux quantum h/2e.)

It must be emphasized that *all* that these experiments show is (a) that everything is consistent with the continued validity of QM at this level, and (b) that *if* the experiments are interpreted according to the QM formalism, then the values of D realized are large. They do *not* prove that macrorealism at this level is excluded.

5. Prospects for Experimental Tests of Macrorealism versus QM

Any discussion of the quantum measurement problem in the 1990s needs to take account of a fact that, I suspect, would have greatly surprised the founders of the subject, namely that there seem to remain no obviously insuperable barriers against testing the theory against realism at a level which may or may not be regarded as "macroscopic" in the fullest sense of the word (no doubt that is a matter of taste) but that is at any rate many orders of magnitude removed from what could have been reasonably imagined in the early days.

Here is one experiment one would ideally like to do (certainly not the only possible one, but the formal analysis and/or the experimental realization seem rather simpler than in some other cases). We find a *single* macroscopic system (*not* an ensemble in the usual sense) characterized, inter alia, by a macroscopic variable that, if we believe the predictions of QM, can oscillate between two regions corresponding to reasonably macroscopically distinct states with some oscillation frequency Δ. This is the macroscopic analog of the NH_3 inversion resonance, and as we shall see there is satisfactory evidence that this situation can be realized in at least one class of experimental systems. We define a suitably coarse-grained variable Q whose eigenvalues ± 1 correspond to the two macroscopically distinct regions, and in a preliminary experiment verify that *when actually measured* the variable Q indeed takes only[3] the values ± 1. The experiment—known in the trade as the search for macroscopic quantum coherence or MQC—then consists in preparing the system repeatedly in the same initial state at time zero, and then measuring, in four different series of runs, the correlations $K(t_i, t_j)$ of Q at two *and only two* (on each series) of the four times t_1, t_2, t_3, t_4. The data obtained are then analyzed in terms of the predictions of (a) quantum mechanics and (b) an arbitrary theory of the macrorealistic type defined below.

To the extent that our system can be treated as an ideal isolated two-state system, the predictions of QM are extremely simple. Independently of the ensemble (that is, of the initial preparation) the correlations $K(t_i, t_j)$ are given by the simple formula

$$K(t_i, t_j)|_{QM} = \cos \Delta(t_j - t_i) \qquad (1)$$

where Δ is the oscillation frequency. It cannot be too strongly emphasized that the prediction (1) applies only under the assumption that the system is undisturbed, either by measurement or by interaction with its environment, for times between t_i and t_j.

We now consider the possibility of describing the experiment by a theory that embodies the property of macroscopic realism as introduced in section 1. Let us define a macrorealistic (MR) theory for this experimental setup by the conjunction of three premises. The first corresponds intuitively to the informal definition given in section 1; in the literature it is sometimes itself called macrorealism, but to avoid possible confusion it may be better to call it "macro-objectivity" (MO):

(A1) The variable Q possesses at any given time t either the value $+1$ or the value -1, irrespective of whether or not it is actually measured at time t. (MO)

In order to derive experimental conclusions from this postulate it is necessary to supplement it with two further ones. The first is the familiar principle of induction as commonly employed, implicitly, in both classical and quantum physics. For our purpose we may write it as:

> (A2) The statistical properties of a given ensemble are determined only by the preparation of that ensemble. (Ind)

The second supplementary assumption may be called "noninvasive measurability" (NIM):

> (A3) It is in principle possible to make at least the first of any two measurements of Q in such a way that the ensemble is not disturbed.

I return below to the question of the status and plausibility of postulate (A3) within an MR theory.

Given postulates (A1–3), it is straightforward to show that we can derive for the quantities $K(t_i, t_j)$ exactly the same sets of inequalities as are derived in the Bell-EPR context, with the t_j's replacing the polarizer settings, e.g., the CHSH inequality (Clauser, Horne, Shimony, and Holt 1969)

$$K(t_1,t_2) + K(t_2,t_3) + K(t_3,t_4) - K(t_1,t_4) \leq 2. \tag{2}$$

Moreover, it is clear that with the choice $t_2 - t_1 = t_3 - t_2 = t_4 - t_3 = \pi/4\Delta$ the inequality (2) is violated by the quantum-mechanical prediction (1). Thus, if the idealized experiment described could be done, we could as it were force Nature to choose directly, at this level, between quantum mechanics and macrorealism.

The questions that can be raised concerning the MQC experiment as just described fall into two main categories, corresponding roughly to (a) can it be done? and (b) what would it show?

Can it be done? Or rather, can an experiment be done that is sufficiently close to the idealized one described to have essentially the same implications? The first question is whether suitable systems, involving a macroscopic variable that in effect takes two and only two distinct values corresponding to macroscopically distinct states, can be found. To this the answer is almost certainly yes: existing experiments with single rf SQUID rings demonstrate beyond reasonable doubt that where biased with an appropriate external flux they behave precisely as predicted by theory, in that the effective potential seen by the relevant macroscopic variable, in this case the total trapped flux or equivalently the circulating current, has the classical double-well form of the NH_3 molecule. The two macroscopically distinct states in this case correspond to currents of the order of 1–10 μA circulating clockwise or counterclockwise, and existing experiments show that the current, when measured, always takes one value or the other.

Moreover, while no experiment has (at least in my opinion) yet observed underdamped NH_3-type oscillations, there is considerable evidence that processes of QM tunneling through the barrier separating the two wells do occur, and that the effect of dissipation on this process is correctly predicted by the theory developed in the past few years (see section 4).

It should be emphasized that even if the experiments of Awschalom et al. (1992) on magnetic biomolecules have indeed observed NH_3-type oscillations of the magnetization in this system, it is not suitable for a test of QM versus MR. What is observed in these experiments (and is likely to be observed for the foreseeable future) is only the average magnetization of the ensemble. Since the eigenvalues of this quantity are not restricted to the values ± 1, theories of the MR type put no useful constraints on the experimentally observed correlations; postulate (A1) cannot be required in the form stated, and the much weaker form that is actually appropriate allows the derivation of no practically useful inequalities.

A more critical question is, Will decoherence render the experiment unable to discriminate between QM and MR? The point is that QM predicts the behavior described by equation (1) only for the case of an ideal, isolated two-level system. Although behavior approximating this is indeed observed (indirectly) in a few microscopic systems such as the NH_3 molecule, even at the microscopic level it is easily washed out by the environment (as happens in many electron transfer reactions in chemical physics); in the light of the extensive decoherence literature it would be natural to suspect that these effects would be overwhelming at the level of a macroscopic system such as an rf SQUID.

Remarkably, it looks as if this conclusion may be much too pessimistic. As I have already emphasized, over the past fifteen years condensed-matter physicists have developed a scheme for handling those effects of interaction of the relevant variable with the environment that in quantum measurement theory are known as decoherence; and this scheme (which, as indicated, is based not on the postulation of a microscopic Hamiltonian but on relating the predicted quantum behavior to that in the semiclassical regime, which is often directly observable experimentally) seems to have been successful, even surprisingly so, in accounting quantitatively for the behavior observed in the fully quantum regime, e.g., in the rate of MQT out of a metastable well. If we assume the scheme to be generally valid (and, of course, that QM will continue to work!), then we can use it to assess the effects of decoherence, and it turns out that they are not so severe that there is not a reasonable prospect of getting around them. In particular, although we would not expect that the prediction (1) is *exactly* fulfilled in any real-life experiment, there seems no reason to doubt that it could be approximated closely enough that the inequality (2) will still be violated by a statistically significant margin.

Indeed, if the experiment when first tried does not "work," I suspect it will be less because of the intrinsic decoherence due to the material environment than to the difficulties associated with the necessity of switching one's measuring apparatus "on" when needed and "off" for the rest of the time. This, however, is a technical problem and as far as I can see there are no ineluctable a priori reasons why it cannot be overcome.

What would the experiment show? Suppose, first, that it comes out *against* the QM predictions. Then I confidently anticipate that most people's reaction would be that there is something wrong with the experiment and/or with the techniques that we have used to apply QM to this case. In particular, all the loopholes in the arguments that we have used to relate the effects of decoherence to the observed classical dissipation would no doubt rightly come under the microscope. I personally believe that given the seamless whole nature of the quantum formalism, it would require something of a pathological conspiracy for those techniques to have given us the rather surprisingly good results they have so far and then to let us down in the "crunch," but certainly this possibility would have to be explored and, ideally, refuted by further theoretical considerations. If we suppose for the sake of argument that this is done, and the resulting QM theory still badly fails to fit the data, then one possible resolution would be that there is associated with the resulting reduction process a timescale (analogous to that of the GRWP theory) somewhere between the scales of the order of 10^{-12} secs, which are operative in the MQT experiments and most others which seem to show QM still working at this scale of complexity, and the much larger timescales ($\sim 10^{-6}$ secs), which should be relevant to the MQC experiment. This would give at least some clue as to the direction in which we should look for such a non-QM theory. Incidentally, it seems unlikely that the GRWP theory itself, when applied in its present form to the SQUID experiment, would predict any substantial deviation from QM, because the two macroscopically distinct states occurring in the experiment differ mainly in the direction of the circulating current (and the associated magnetic field); the difference in the position of the center of mass, which is the crucial variable in the GRWP theory as currently formulated, is extremely small. However, a suitably generalized version of the theory might well predict a substantial difference.

Now suppose that—as I imagine most physicists would bet—the experiment comes out according to the predictions of QM, and moreover the latter are such that they violate the MR inequality (2). What exactly can we infer from this that we did not already know? My own reaction would be simply that while we have always known that QM, if we believe it, says that our everyday common-sense notions about the macroscopic world must be wrong, we have now actually directly demonstrated them to be wrong. Can we be more specific than that? The literature of the past few years has included a considerable amount of discussion as to whether, if the QM result is found, we would have to reject postulate (A1) (macro-objectivity) or could get away, instead, with rejection only of (A3) (non-invasive measurability). No one has to my knowledge seriously suggested that we abandon (A2) (induction). Frankly I am not sure that this question is really very meaningful. The everyday language that we use to describe the macroscopic world is based on a whole complex of implicit, mutually interlocking assumptions, so that once the complex as a whole is seen to fail it may not make much sense to ask which particular assumption is at fault. I am not sure, myself, that I could give a lot of meaning to postulate (A1) under conditions where I had to admit that (A3) fails. To amplify this comment, let me imagine that we can measure Q (for the first of t_i, t_j at least) by an "ideal-negative-result" (INR) technique: we arrange that the system interacts physically with the measuring apparatus only if it is in *one* of its two states, while in the other no interaction takes place. (This is the analog of a Young's slits experiment in which we shine a flashlight only on one of the slits.) It is easily shown that a series of experiments of this type, *in which we throw away all runs in which the measurement apparatus reacts*, is sufficient to measure the correlation $K(t_i, t_j)$. Moreover, we can do a supplementary series of experiments in which we choose t_i to be a time such that $Q(t_i)$ is known (and verified) to be always $+1$, and we compare the statistics of $Q(t_j)$ $(t_j > t_i)$ under the conditions (a) that Q is not measured at any time prior to t_j, and (b) that Q is measured by an INR technique at t_i (that is, the apparatus is set to react only if $Q = -1$). Assuming that the two ensembles give the same result for the statistics of $Q(t_j)$ (as QM predicts), any attempt to maintain (A1) by discarding (A3) would then seem to force on us the following thesis: Whenever QM tells us that Q has *for sure* the value $+1$, then an INR measurement is noninvasive; but in the more general case, despite the fact that Q actually has one of the values ± 1 (postulate A1) the corresponding INR measurement is invasive. Under these circumstances I am not sure what meaning I should attach to the words "actually has," and I suspect that the whole debate becomes devoid of empirical content. The important point is surely that "macroscopic common sense" has been seen, dramatically, to fail.

Whatever future research in this direction may or may not show, I hope to have demonstrated (a) that, at least under some—admittedly specially engineered—conditions, the effects of decoherence have been grossly overestimated in much of the existing literature, and (b) that irrespective of whether or not the MQC experiment turns out to be technically feasible in the next few years, it can at least serve as a useful testbed for alternatives to QM and/or different interpretations of the quantum formalism.

Notes

This work was supported by the MacArthur Chair endowed by the John D. and Catherine T. MacArthur Foundation at the University of Illinois.

1. The lower boundary of the spatial scale on which QM has been actually "verified" to work is presumably that corresponding to current maximum accelerator energies, i.e., about 10^{-4} fm, considerably less than halfway from the atomic to the Planck scale.

2. After this was written there has appeared a remarkable paper (M. Brune et al., Phys. Rev. Letters 77, 4887 (1996)) which presents direct evidence for a state of such a cavity with D of order 10.

3. In principle we should expect very occasionally to find the system "in transit" (under the barrier, see continued discussion in the chapter), and a completely rigorous analysis must allow for this.

References

Awschalom, D. D., J. F. Smyth, G. Grinstein, D. P. di Vincenzo, and D. Loss. 1992. *Phys. Rev. Lett.* 68, 3092.

Bell, J. S. 1964. *Physics* 1, 195.

Caldeira, A. O., and A. J. Leggett. 1983. *Ann. Phys.* 149, 347: erratum, ibid. 153, 445 (1984).

Chechetkin, V. R. 1976. *Fiz. Nizk. Temp.* 2, 434: trans. *Sov. J. Low Temp. Phys.* 2, 215.

———. 1982. *Fiz. Nizk. Temp.* 8, 41: trans. *Sov. J. Low Temp. Phys.* 8, 9.

Clarke, J., A. N. Cleland, M. H. Devoret, D. Esteve, and J. M. Martinis. 1988. *Science* 239, 992.

Clauser, J. F., M. A. Horne, A. Shimony, and R. A. Holt. 1969. *Phys. Rev. Lett.* 23, 880.

Ellis, J., J. S. Hagelin, D. V. Nanopoulos, and M. Srednicki. 1984. *Nucl. Phys.* B 241, 381.

Garg, A. 1993. *Phys. Rev. Lett.* 71, 249.

Greenberger, D. M., M. Horne, and A. Zeilinger. 1989. In *Bell's Theorem, Quantum Theory, and Conceptions of the Universe*, ed. M. Kafatos. Dordrecht: Kluwer Academic, 73.

Griffiths, R. B. 1984. *J. Stat. Phys.* 36, 219.

Gunther, L., and B. Barbara, eds. 1995. *Quantum Tunneling of the Magnetization* Proc. 1994 Chichilianne NATO Workshop, Kluwer Academic.

Leggett, A. J. 1980. *Prog. Theor. Phys.*, suppl. 69, 80.

———. 1984. *Phys. Rev.* B 30, 1208.

———. 1990. In *Applications of Statistical and Field Theory Methods to Condensed Matter* (Proc. 1989 NATO Summer School, Evora, Portugal), ed. D. Baeriswyl, A. R. Bishop, and J. Carmelo. Plenum Press.

Pearle, P. 1989. *Phys. Rev.* A, 39, 2277.

Penrose, R. 1986. In *Quantum Concepts in Space and Time*, ed. R. Penrose and C. J. Isham. Oxford: Oxford University Press, 129–46.
Pines, A. 1988. In Proc. Int. School of Physics "Enrico Fermi," course C, ed. B. Maraviglia, North-Holland.
Rouse R., S. Han, and J. E. Lukens. 1995. *Phys. Rev. Lett.* 75, 1614.
Unruh, W. G. 1995. "False Decoherence." Preprint.
Yip, S. K. 1984. *Phys. Lett.* A 105, 66.
Zurek, W. H. 1991. *Phys. Today*, Oct. 1991, 36.

Abner Shimony

Comments on Leggett's "Macroscopic Realism"

Anthony Leggett's studies of quantum effects in macroscopic systems have stimulated the investigation of fascinating phenomena and will almost certainly lead to the discovery of others. Since the contemplation of phenomena with pleasure is an essential part of the enterprise of natural philosophy, we philosophers thank him for introducing us to macroscopic tunneling of magnetic flux trapped in a SQUID ring, for the prospect of seeing the resonance effect that he calls "macroscopic quantum coherence," and for his discussions of numerous related effects in Josephson junctions, magnetic grains, and macromolecules.

In addition, Leggett's program is extraordinarily valuable for deepening our analysis of the foundations of quantum mechanics. If quantum mechanics is a physical theory of unlimited validity, not merely as an instrument for systematizing and predicting experimental outcomes but as a characterization of the properties of physical things themselves, then there is a well-known anomaly implicit in it. The quantum mechanical superposition principle would apply to states of macroscopic as well as microscopic systems, and there are situations, under appropriate initial conditions, where the measurement of a property of a microscopic system would imply a superposition of macroscopically different states of the measuring apparatus. How then can there be a definite outcome of the measurement procedure? This problem is commonly called the "measurement problem," the "problem of the actualization of potentialities," and the "Schrödinger cat problem." One of the most attractive avenues for dealing with it is summarized by Leggett as follows: "that QM as presently conceived is *not* the whole truth about the world, but that at some level of physical scale, complexity, or degree of organization there come into play new and currently unknown physical laws whose nature is such as to guarantee that it is never necessary to describe the state of the world, even formally, in terms of a quantum superposition of macroscopically distinct states" (section 1). This thesis—which I should confess from the start seems to me very attractive—is called "macroscopic realism" early in Leggett's chapter and later called "macro-objectivity" (section 5). The core of Leggett's program is to assess this thesis.

There are several noteworthy and admirable features of Leggett's conduct of his program. First, he maintains an independent mind in the face of a heavy accumulation of accepted lore. For example, in section 3 he rejects the proposition that superconductors and superfluids provide instances of macroscopic superpositions; he rejects Bohr's idea that the behavior of a macroscopic variable is so far into the quasi-classical limit that the predictions of QM agree with those of classical physics; and he objects to the proposition that it is experimentally impossible to distinguish between a superposition of macroscopic states and a mixture. A second admirable feature—hardly separable from the first—is his profound and detailed theoretical knowledge of condensed matter. I found the following passage so illuminating that I wish to quote it here:

> The fact that macroscopic bodies have incredibly closely spaced energy levels—sometimes cited as killing any possibility of QIMDS—is largely irrelevant, since most of them are effectively decoupled from the variable of interest. . . . Uncertainty about the microscopic initial conditions is similarly much less important than it has often been taken to be, since, again, it usually does not affect the motion of the relevant macroscopic variable appreciably. . . . Most important, even quite strong coupling to the environment does not automatically lead to the destruction of QIMDS. (Section 4)

(QIMDS is the unpronounceable acronym for "quantum interference between macroscopically distinct states," a logopathy that Leggett committed in spite of his fine literary and philological training.) A third admirable feature is that he has imagination to propose performable experiments that can exhibit the presence or absence of macroscopic superposition. And finally, he is open-minded regarding the experimental outcome, equably considering the scientific and philosophical consequences of whatever the experimenters find.

After these general remarks about the character of Leggett's program I shall focus my attention on the "macroscopic quantum coherence" (MQC) experiment. Why does he feel that this difficult experiment is needed in order to supplement the quite well established phenomenon of "quantum macroscopic tunneling" (QMT)? Why is a temporal version of Bell's inequality a legitimate tool for analyzing MQC? And what is the most promising avenue for solving the measurement problem if (as Leggett seems to expect) his proposed experiment turns out to confirm the existence of MQC?

In a standard quantum-mechanical treatment of tunneling (for example, Schiff 1955, 72), the stationary solution of the time-dependent Schrödinger equation is a superposition of one term describing the particle as confined to a region from which classically it could not escape and another term describing the particle as occupying a classically inaccessible region. The rate of tunneling is calculated from the coefficients of the two terms, but that is a quantitative matter that need not concern us here. The relevant point is the qualitative one that the phenomenon of tunneling is quantum-mechanically explained as the superposition of two terms which differ radically with regard to some physical property. The quantum-mechanical treatment of tunneling of magnetic flux (de Bruyn Ouboter 1984, and references in that work) follows the generic pattern, but with the novel feature that the superposed terms differ with respect to magnetic flux and therefore with respect to current in the Josephson junction, and the difference in the current is macroscopic. The situation can be summarized with acronyms: if the phenomenon is explained by QM, then it is an instance of QIMDS. But what Leggett is saying at the end of section 4 is that a non-quantum-mechanical explanation has not been ruled out, just as a hidden-variables interpretation of spin phenomena for a single spin–$\frac{1}{2}$ particle is not ruled out (Bell 1966; Kochen and Specker 1967).

In the proposed MQC experiment the temporal correlation function $K(t_1,t_2)$ will satisfy a Bell-type inequality—inequality (2)—if the three postulates A1, A2, and A3 are satisfied, these three together constituting the definition of a "macrorealistic theory" (section 5). Just as the violation of an analogous Bell-type inequality for a pair of correlated photons (in an experiment where the well-known "communication" and "detection" loopholes are blocked) would rule out a local hidden-variables explanation of the correlations, leaving the standard explanation of QM without a well-articulated rival (Clauser and Shimony 1978; Selleri 1990), so also would the violation of inequality (2) in the experiment proposed by Leggett leave no well-articulated rival to QM concerning the behavior of trapped flux in a Josephson junction, provided of course that there are no loopholes. But is there not a gaping loophole, much worse than in the photon correlation experiments? Regarding the experiment of Aspect et al. (1982), one might doubt that the communication loophole has been blocked because the polarizer switches change from off to on with a regular period of 20 nanoseconds, and clever hidden variables in one wing of the experiment could thereby infer the orientation of the polarization analyzer in the other wing. All the more skeptical should a critic of Leggett's experiment be, since in it one and the same system is observed at two different times t_1 and t_2, and it is far from obvious that the result (or probability of the result) at t_2 is independent of the time of first intervention t_1 and the result found at that time. A skeptic would interpret a violation of inequality (2) not as a triumph of QM over macrorealism, but only as an indication that macrorealism is compatible with instability, so that small interventions can cause large changes of outcome. This skeptical doubt proposes to save macro-objectivity (postulate A1) by jettisoning noninvasive measurability (postulate A3). Leggett argues against this skepticism by saying "I am not sure, myself, that I could give a lot of meaning to postulate (A1) under conditions where I had to admit that (A3) fails," and he goes on to elaborate his argument by considering "ideal negative-result" (INR) techniques. A few remarks may be useful as supplements to Leggett's argument.

The first remark is that Leggett's postulate (A3) obviously suffices to derive the factorization condition

$$E_\lambda(t_1,t_2) = E_\lambda(t_1)E_\lambda(t_2),$$

where $E_\lambda(t)$ is the expectation value of Q at time t if λ is the complete state of the Josephson junction, and $E_\lambda(t_1,t_2)$ is the expectation value of the products of Q at t_1 and t_2 if λ is the complete state of the junction. Except for changes of notation the factorization condition is the same as what Bell calls the "locality" condition in his pioneering papers (1964, 1971). It follows from the factorization condition and the fact that Q has absolute value equal to or less than unity that if the correlation function $K(t_1,t_2)$ is defined as

$$K(t_1,t_2) = \int d\rho \; E_\lambda(t_1,t_2)$$

where ρ is an arbitrary probability distribution over the space of complete states, then K satisfies inequality (2). It may be surprising to those who have not worked much with Bell's inequalities that the factorization condition precludes the cosinusoidal form of $K(t_1,t_2)$ of Leggett's equality (1), but once one follows a standard derivation of inequality (2), one sees that the surprise must be accepted. To derive inequality (2) it is not necessary to use Leggett's postulate (A1), which states that Q has a definite value when the complete state is specified, but of course (A1) is a consistent assumption.

The second remark is that postulate (A3), noninvasive measurability, is much more plausible when one is dealing with a macroscopic rather than a microscopic system. The possession of a property by a microscopic system could be unstable in the sense that a minimum disturbance would suffice to change its value. By contrast, the two possible values of the variable Q in Leggett's experiment, +1 and −1, correspond to macroscopically different currents in the ring of the Josephson junction and hence to macroscopically different amounts of a conserved quantity, angular momentum.

It is reasonable that a measurement that transfers very little angular momentum would suffice to discriminate between Q = +1 and Q = −1, and hence would leave Q unchanged.

The third remark is that QM is able to predict a violation of inequality (2) because it does not satisfy (A3), and its failure to satisfy (A3) is closely linked to its failure to satisfy (A1). If the concept of state in QM is interpreted in terms of potentiality, as suggested by Heisenberg (1962, 185), then a superposition of two states, in one of which Q has the value +1 and in the other Q has the value −1, would violate (A1), since Q would then have the status of a potentiality and not an actuality. If Q is measured and a definite result is obtained, then there is a change of state of the Josephson junction, contrary to postulate (A3). But the invasion does not entail a change of Q from +1 to −1 or vice versa, which is problematic, as argued in the preceding paragraph. The invasion has a character without precedent in classical physics: it is the actualization of a potentiality. I find this novel type of invasion of a system—actualization of a potentiality—to be intellectually very satisfying and appropriate to the phenomenon that Leggett expects will be found. What keeps it from being entirely satisfactory, however, is the mysteriousness of the actualization of potentialities when the fundamental physical dynamics is linear.

If, as Leggett expects, the macroscopic quantum coherence experiment will yield results violating inequality (2) and agreeing with the predictions of QM, and if, furthermore, macro-objectivity cannot be salvaged by jettisoning noninvasive measurability, then the measurement problem cannot be solved by the exclusion of macroscopic systems in general from the domain of validity of the superposition principle. Consequently, an avenue for solving the measurement problem, which a priori is very attractive, will have been blocked. Where then would we find ourselves? We have strong empirical evidence that the superposition principle holds for microscopic systems; we have direct introspective evidence that when we read a measuring device that has been used to measure an initially indefinite quantity of a microscopic system we obtain a definite reading; but how are these two termini to be connected and reconciled? When a tunnel was dug through Mt. Blanc, excavations were initiated from both the Swiss and Italian sides, and by a triumph of engineering the two excavations neatly met each other under the mountain. As a research worker in the foundations of quantum mechanics I sometimes have the sickening feeling of belonging to an engineering team whose excavations from two sides of a mountain have ignominiously missed each other. In particular, if the MQC experiment comes out as Leggett anticipates, exhibiting coherence, then the macroscopic character of the measuring apparatus cannot be relied upon to serve as the locus of the meeting.

An idealized formulation of the quantum-mechanical measurement process presents an array of possible solutions to the measurement problem. Suppose that the micro-object of interest is initially in the state $Q = \Sigma c_i u_i$, where the u_i are orthonormal eigenvectors of a quantity of interest of the microscopic object. Suppose that this object interacts with an apparatus that is initially in a "neutral" state v_0 and that the interaction Hamiltonian is such that if the object is initially in u_i then at time t the object plus apparatus is in $u_i \otimes v_i$, where the v_i are eigenstates of a macroscopic variable of the apparatus. Inevitably other systems become entangled with the object and the apparatus: notably, the physical environment, the space-time metric, the molecules of the observer's sensory and cognitive faculties, and finally the observer's psyche. Because of the unitarity of the time-evolution operator of QM, interactions bring about an entangled state of the great composite system consisting of all these components, and for simplicity we shall suppose that the entangled state can be expressed by a single sum of orthogonal "branches":

$$\Psi = \sum c_i u_i \otimes v_i \otimes w_i \otimes x_i \otimes y_i \otimes z_i,$$

where the w_i are orthogonal states of the environment, x_i of the space-time field, y_i of the molecules of the observer's body, and z_i of the observer's psyche. In order to have a definite perceptual result, the "chain of statistical correlations" in the state Ψ must be "cut" (terminology of London and Bauer [1939], section 11), producing the nonlinear transition or "reduction"

$$\Psi \rightarrow u_k \otimes v_k \otimes w_k \otimes x_k \otimes y_k \otimes z_k$$

for some index k. If one accepts this way of posing the measurement problem (modified unavoidably if idealizations are removed), and does not attempt to solve the problem by a hidden-variables theory, a decoherence theory, or a related strategy, then one must ask where and how the reduction occurs. If the MQC experiment comes out as Leggett expects, then one cannot claim that the locus of the reduction is the apparatus on the ground of its being the first macroscopic system in the chain of correlations. London and Bauer (1939) proposed instead that the locus of the reduction is the psyche. Károlyházy et al. (1986), Penrose (1986), and others have proposed that space-time is the locus.

I am not sure that anyone has proposed the environment as the locus of reduction, even though interaction of a macroscopic system with the environment is a central aspect of decoherence theories; but these retain the unitary dynamics of standard QM and therefore supply only "as if" reduction rather than the real thing (Griffiths 1984; Gell-Mann and Hartle 1993; Omnès 1992; and others). A possibility that seems to me largely to have been neglected in the literature on the measurement problem (an exception being Percival 1992) is that the locus of reduction is the macromolecules of the sensory and cognitive faculties.

A concrete example will provide some motivation for my conjecture. The photoreceptor protein of the rod cells, rhodopsin, is known to absorb a photon and initiate a biochemical cascade that eventually produces a macroscopic pulse in the optic nerve (Stryer 1987). The two components in rhodopsin are retinal, which can absorb a photon, and opsin, which acts as an enzyme that effects the binding of about five hundred mediating molecules when it is triggered by the excited retinal. (There are further multiplication factors later in the cascade, but I shall restrict my attention to the earlier stages.) In the resting state of retinal, hydrogen atoms attached to the eleventh and twelfth carbon atoms lie on the same side of the carbon backbone (so that the conformation is called *cis*), and this arrangement causes the backbone to bend. There is a potential barrier between the *cis* conformation and the *trans* conformation, in which the two hydrogen atoms mentioned are on opposite sides of the backbone from each other. Spontaneous tunneling between the *cis* and the *trans* conformations almost never occurs. But when retinal in the *cis* conformation absorbs a photon, it acquires sufficient energy for a rotation to occur between the eleventh and twelfth carbon atoms, so that the *trans* conformation is achieved. Now we raise a question analogous to the one Schrödinger posed concerning his famous cat: what if the unitary dynamics of evolution of the photon and the retinal produces a superposition of the *cis* and the *trans* conformations? That would be analogous to a superposition of the living and the dead states of the cat. Would not such a superposition produce an indefiniteness of seeing or not seeing a visual flash, unless, of course, a reduction occurred further along the pathway from the optic nerve to the brain to the psyche? My conjecture is that the reduction occurs at the retinal molecule itself: that there is a superselection rule operative which prevents a superposition of molecular conformations as different as *cis* and *trans* from occurring in nature. A general superselection rule of this kind would have the desirable consequence that in intracellular processes a molecular "switch" is never in a superposition of "off" and "on," since these correspond to different conformations.

An objection will immediately be raised on the basis of Leggett's work: if superpositions of distinct macroscopic states of macroscopic systems can occur, as a positive outcome of the MQC experiment would show, then *a fortiori* nature should not deny the applicability of the superposition principle to a biological molecule, which is much smaller than macroscopic. An answer could be that the line of delimitation between the applicability and nonapplicability of the superposition principle must be something more subtle than a certain size or a certain number of degrees of freedom. Structure is also important and may be crucial. Considerations of structure may be responsible for the (hypothetical) fact that certain states of macroscopic systems are superposable while a superselection rule governs certain molecular conformations.

One of the attractive features of my conjecture is its testability, or at least the testability of something like it. Organic molecules with different handedness, like dextrose and levulose, are found in nature, but superpositions of different handedness, such as the positive and negative eigenstates of the spatial inversion (or parity) operator, do not seem to be found. Is this fact due to a dynamical instability, which would cause a parity eigenstate to decay rapidly into a mixture of dextrose and levulose, or is it due to a genuine superselection rule (Primas 1981, 12)? Several proposals have investigated this important question experimentally (Quack 1989; Silbey and Harris 1989). Suppose the answer turns out to be the latter—a genuine limitation on the superposition principle. Then there would be some plausibility in the conjecture that the *cis* and *trans* conformations of retinal cannot be superposed.

Suppose, on the other hand, that a superposition of dextrose and levulose is momentarily detected. Does that mean that we should have to worry seriously about biomolecular analogues of Schrödinger's cat? Maybe not. We might conjecture that such superpositions occur only when the molecules are well isolated from other systems, for instance when a molecular beam is prepared, but that the biological environment of a biomolecule *in vivo* would prevent superposition of different states of handedness or of different conformations. In other words, my conjecture that the locus of reduction of a superposition is the biomolecule would be modified, and the locus would be *the molecule in its environment*. The modified conjecture begins to sound like decoherence theory. However, I mean something different: not the negligibility of off-diagonal elements in a statistical matrix or the indistinguishability of a superposition from an appropriate mixture, but rather the literal actualization of a potentiality. Thus the modified conjecture is closer to stochastic modifications of quantum dynamics, such as those of Ghirardi, Rimini, and Weber (1986), Diósi (1989), Gisin (1989), and Pearle (1989). None of these radical modifications of quantum dynamics has been worked out for systems as complex as the biological environment. An immense amount of work would be required to adapt these stochastic theories to biophysics.

•

The research for this paper was supported in part by the National Science Foundation, grant no. PHY93–21991.

References

Aspect, A., J. Dalibard, and G. Roger. 1982. *Phys. Rev. Lett.* 49, 1804–7.
Bell, J. S. 1964. *Physics* 1, 195. Reprinted in J. S. Bell, *Speakable and Unspeakable in Quantum Mechanics*. Cambridge: Cambridge University Press, 1987.
———. 1966. *Rev. Mod. Phys.* 38, 447. Reprinted in J. S. Bell, *Speakable and Unspeakable in Quantum Mechanics*. Cambridge: Cambridge University Press, 1987.
———. 1971. In *Foundations of Quantum Mechanics*, ed. B. d'Espagnat. New York: Academic Press, 171. Reprinted in Bell 1987.
de Bruyn Ouboter, R. 1984. In *Foundations of Quantum Mechanics in the Light of New Technology*, ed. S. Kamefuchi et al. Tokyo: Japan Physical Society, 83.
Clauser, J. F., and A. Shimony. 1978. *Rep. Progr. in Phys.* 41, 1881–1927.
Diósi, L. 1989. *Phys. Rev.* A 40, 1165–74.
Gell-Mann, M., and J. B. Hartle. 1993. *Phys. Rev.* D 47, 3345–82.
Ghirardi, G.-C., A. Rimini, and T. Weber. 1986. *Phys. Rev. D* 34, 470–91.
Gisin, N. 1989. *Helv. Phys. Acta* 62, 363–71.
Griffiths, R. B. 1984. *J. Stat. Phys.* 36, 219–72.
Heisenberg, W. 1962. *Physics and Philosophy* (New York: Harper).
Károlyházy, F., A. Frenkel, and B. Lukács. 1986. In *Quantum Concepts of Space and Time*, ed. R. Penrose and C. Isham. Oxford: Oxford University Press. 109–28.
Kochen, S., and E. P. Specker. 1967. *J. Math. and Mech.* 17, 59–87.
London, F., and E. Bauer. 1983. In *Quantum Theory and Measurement*, ed. J. A. Wheeler and W. H. Zurek. Princeton, N.J.: Princeton University Press. 217–59. Translated from *La théorie de l'observation en mécanique quantique* (Paris: Hermann, 1939).
Omnès, R. 1992. *Rev. Mod. Phys.* 64, 339–82.
Pearle, P. 1989. *Phys. Rev.* A 39, 2227–89.
Penrose, R. 1986. In *Quantum Concepts of Space and Time*, ed. R. Penrose and C. Isham. Oxford: Oxford University Press. 129.
Percival, I. 1992. *Nature* 351, 357.
Primas, H. 1981. *Chemistry, Quantum Mechanics, and Reductionism.* Berlin: Springer-Verlag.
Quack, M. 1989. *Angew. Chem. Int. Ed. Engl.* 28, 571–86.
Schiff, L. 1955. *Quantum Mechanics*, 2d ed. New York: McGraw-Hill.
Selleri, F. 1990. *Quantum Paradoxes and Quantum Reality*. Dordrecht: Kluwer.
Silbey, R., and R. A. Harris. 1989. *J. Phys. Chem.* 93, 7062–71.
Stryer, L. 1987. *Scientific American* 257 (July), 42–50.

Jeffrey Bub, Rob Clifton, and Bradley Monton

The Bare Theory Has No Clothes

1. What Makes the Bare Theory So Seductive?

Consider quantum theory without the collapse postulate. No experiment has ever disconfirmed its statistical predictions. So one might reasonably bet that no experiment ever will. If not, every possible state of anything in the universe, from the spin states of electrons to the states of consciousness of sentient beings, must evolve in time according to Schrödinger's equation. A theoretician's paradise . . . except for that nagging problem of how to make sense of what the theory says will happen when we try to confirm its statistical predictions.

The problem is worth rehearsing. Suppose m is a device that reliably records whether an electron e's spin is *up* or *down* along some specified direction (for simplicity: without disturbing that spin). The unitary evolution characterizing the $m + e$ interaction will therefore map:

$$
\begin{aligned}
|up\rangle_e \, |\text{'?'}\rangle_m &\rightarrow |up\rangle_e \, |\text{'}up\text{'}\rangle_m, \\
|down\rangle_e \, |\text{'?'}\rangle_m &\rightarrow |down\rangle_e \, |\text{'}down\text{'}\rangle_m,
\end{aligned}
\tag{1}
$$

where $|\text{'?'}\rangle_m$ is the ready-to-measure state of m, and $|\text{'}up\text{'}\rangle_m$ and $|\text{'}down\text{'}\rangle_m$ are two orthogonal recording states of m that distinguish whether e's spin is *up* or *down* in the direction of measurement. (The quotation marks signify that, for present purposes, we can remain agnostic about precisely what physical quantity of m these three states are eigenstates of.) If m is set up to do its job when e is in some arbitrary superposition of *up* and *down* spin states:

$$
c_1 \, |up\rangle_e + c_2 \, |down\rangle_e, \tag{2}
$$

the above evolutions and the linearity of the Schrödinger equation entail that the final state of $m + e$ will be:

$$
c_1 \, |up\rangle_e \, |\text{'}up\text{'}\rangle_m + c_2 \, |down\rangle_e \, |\text{'}down\text{'}\rangle_m. \tag{3}
$$

Quantum theory dictates that, on looking to see what recording property *m* has, there is a chance of $|c_1|^2$ that an experimenter, let's call her Eve, will see '*up*,' and a chance of $|c_2|^2$ that she will see '*down*.'

Now if the Schrödinger equation is universally valid and Eve is a competent observer, we should be able to model the acquisition of her belief about the spin state of *e* (which she acquires through her belief about *m*'s recording property) by a unitary mapping of the same general form as (1). Then Eve's looking at *m* will generate the state:

$$\begin{aligned} &c_1 \; |up\rangle_e \; |\text{`}up\text{'}\rangle_m \; |\text{`Believes } e\text{-spin } up\text{'}\rangle_{\text{Eve}} \; + \\ &c_2 \; |down\rangle_e \; |\text{`}down\text{'}\rangle_m \; |\text{`Believes } e\text{-spin } down\text{'}\rangle_{\text{Eve}}, \end{aligned} \tag{4}$$

where the two states of Eve in this superposition[1] are vectors lying in distinct eigenspaces of whatever physical quantity of Eve's brain it is that records information gathered through her senses (again, the quotation marks allow us to remain agnostic about what physical quantity that is).

At this point, for the statistical predictions of quantum mechanics to have something to refer to, it seems that one must say that, although (4) is the full quantum state of $e + m +$ Eve, in any particular case when Eve looks at *m*, just one of (4)'s terms represents what Eve will actually come to believe about *e*'s spin. Indeed, statistical predictions aside, something like this *must* be said if we are to reconcile the theory with the fact that experimenters like Eve always take themselves to have definite beliefs about what their measurement devices indicate.

But the trouble is that the standard way of thinking about superpositions in quantum mechanics prohibits attributing a definite value to any observable of a system whose eigenstates are superposed by the system's quantum state, as Eve's belief states are in (4). To say anything more than that by way of underwriting the definiteness of Eve's *e*-spin belief with hidden variables, worlds, minds, or what-not is standardly taken to be adding to the theory, and that's standardly taken to be a bad thing to do. So it appears that standard thinkers—at least those who want to uphold the universal validity of the Schrödinger equation without introducing the collapse postulate—have backed themselves into a corner on the issue of whether experimenters like Eve have definite beliefs about their measurement outcomes (not to mention the statistics of those outcomes).

But maybe not. David Albert (1992, 116–19) has suggested that if the issue is simply whether it is possible to recover our everyday sense that experimenters hold definite beliefs about the things with which they interact, then that *can* be done with standard thinking using what Albert calls the Bare theory.

The Bare theory promises to be just what its name suggests: a theory on which the Schrödinger equation *is* universally valid, the standard way of thinking about superpositions *is* correct, and that is all there is to say. So there are no collapses, variables, worlds, or minds postulated to save the definiteness of Eve's belief in (4).[2] The Bare theory's explanation of why, despite all that, everyone (Eve included) will always *come to believe* that her belief about *e*'s spin is definite, is described by Albert as "amazingly cool."[3]

The explanation goes like this (with minor embellishment). Suppose Eve's partner Adam is determined to figure out whether Eve has a definite belief about *e*'s spin while in state (4). Adam cannot just ask Eve what belief about *e*-spin she has while in state (4), because the linearity of the Schrödinger equation will put her into a superposition of responding to Adam in two different ways, and according to Albert (1992, 117), "it won't be any easier to interpret a 'response' like that than it was to interpret the superposition of brain states in [(4)] that that response was intended to be a *description* of!"

Without automatically inducing a superposition of responses, however, Adam can simply ask Eve: "Don't tell me whether you believe the electron to be [*up*] or you believe it to be [*down*], but tell me merely whether or not *one* of those two is the case; tell me (in other words) merely whether or not you *have* any particular definite belief (not uncertain and not confused and not vague and not superposed) about the value of the [spin] of this electron" (1992, 118). The way to model Eve answering that question within quantum mechanics is, presumably, as follows.

If Eve is an honest and competent reporter of her mental states, then when asked whether she has a definite *e*-spin belief in a state like $|\text{`Believes } e\text{-spin } up\text{'}\rangle_{\text{Eve}}$, she should report Yes. For in that case, the Bare theory predicts that Eve's belief *will* be definite, and we can assume that there is nothing else stopping her from honestly and competently reporting that fact. But then the same goes for how Eve will respond if she were asked the same question by Adam while in the state $|\text{`Believes } e\text{-spin } down\text{'}\rangle_{\text{Eve}}$; namely, she will again respond with a Yes.[4]

It follows that the correct way to model Eve's response to Adam's question about the definiteness of her *e*-spin belief is in terms of a unitary evolution that maps:

$$\begin{aligned} &|\text{`Believes } e\text{-spin } up\text{'}\rangle_{\text{Eve}}\,|\text{`Ready to Answer'}\rangle_{\text{Eve}} \\ &\to |\text{`Believes } e\text{-spin } up\text{'}\rangle_{\text{Eve}}\,|\text{`Yes, I have a definite } e\text{-spin belief'}\rangle_{\text{Eve}}, \\ &|\text{`Believes } e\text{-spin } down\text{'}\rangle_{\text{Eve}}\,|\text{`Ready to Answer'}\rangle_{\text{Eve}} \\ &\to |\text{`Believes } e\text{-spin } down\text{'}\rangle_{\text{Eve}}\,|\text{`Yes, I have a definite } e\text{-spin belief'}\rangle_{\text{Eve}}, \end{aligned} \tag{5}$$

where the two states of Eve we have introduced denote the physical correlates in her brain of her getting ready to answer, and responding Yes to the question.[5]

But now the cool thing is that if (5) models Eve's response to Adam, the linearity of the Schrödinger equation demands that when she responds to his question in superposition (4), the final state of $e + m +$ Eve will necessarily be:

$$c_1 \,|up\rangle_e \,|\text{`}up\text{'}\rangle_m \,|\text{`Believes } e\text{-spin } up\text{'}\rangle_{\text{Eve}} \,|\text{`Yes, I have a definite } e\text{-spin belief'}\rangle_{\text{Eve}} + \quad (6)$$
$$c_2 \,|down\rangle_e \,|\text{`}down\text{'}\rangle_m \,|\text{`Believes } e\text{-spin } down\text{'}\rangle_{\text{Eve}} \,|\text{`Yes, I have a definite } e\text{-spin belief'}\rangle_{\text{Eve}},$$

which is *not* a superposition of Eve giving different answers, but an eigenstate of Eve responding Yes! And so the Bare theory predicts that Yes will definitely be Eve's answer, even when in fact she has no definite state of belief about *e*-spin according to that theory (that is, according to standard thinking about what it means to be in an uncollapsed superposition of belief states). Eve is "apparently going to be radically deceived even about what *her own occurrent mental state* is" (1992, 118). Albert calls a situation like this a situation in which Eve "effectively knows" what *e*'s spin is (1992, 120).

So the cool thing is that we can use the sparse resources of the Bare theory to show why no two people are ever going to believe anything out of the ordinary about the definiteness of each other's beliefs. Indeed, since it could just as well have been Eve herself who introspects about her own beliefs and inquires as to whether they are definite about *e*-spin, the Bare theory predicts that experimenters will never believe anything out of the ordinary about their own beliefs either.

It is rather like the situation before the fall: Both Adam and Eve were naked, but to keep the paradise, God ensured that they knew not that they were naked. In this case, the Schrödinger equation is what keeps the paradise:

> That is: maybe (even if the standard way of thinking about what it means to be in a superposition is the right way of thinking about what it means to be in a superposition) the linear dynamical laws are nonetheless the complete laws of the evolution of the *entire* world, and maybe all the appearances to the contrary (like the appearance that experiments have outcomes, and the appearance that the world doesn't evolve deterministically) turn out to be just the sorts of *delusions* which *those laws themselves* can be shown to *bring on*! (1992, 123)

Of course, paradise in the garden of Eden did not last long. And trouble is lurking in the Bare theoretician's paradise too. In fact, this story has at least two problems. We shall draw out the first problem in the next section, and identify the second, the decisive problem, in the section after that.

Albert gives further Bare theory stories to explain why experimenters take immediately repeated measurements to yield the same determinate result, why different experimenters measuring the same observable take themselves to agree on the outcome, and why experimenters take the measurement result statistics they gather in their laboratories to confirm the statistical predictions of quantum mechanics (1992, 119–23). We believe that those stories have exactly the same problems, and shall indicate why toward the end of this chapter.[6]

2. The Bare Necessities

The first thing to sort out is how much of this story about Eve always taking her beliefs to be definite and responding accordingly is a necessary consequence of the Bare theory. Albert homes in on the correct way to model Eve's response to Adam's question by saying that she, being honest and competent enough to assess her own mental state, would answer Yes if she were, according to the Bare theory, in any quantum state that corresponds to her having a definite *e*-spin belief. Albert then employs that model, defined by equations (5), in a state where the Bare theory blocks Eve from having a definite belief. Finally, he concludes on the basis of the linearity of the Schrödinger equation that, nevertheless, she would still respond Yes. But then it would seem she cannot have been honest and competent after all! Perhaps the correct conclusion is that Albert has stacked the deck in his own favor by not modeling Eve so that she can respond as competently and reliably as possible, whatever her circumstances.

But that conclusion is too quick. It fails to recognize that the Bare theory itself sets definite limits on modeling Eve's response. We summarize these limits in a modest theorem.

Bare Theorem

The following are mutually inconsistent:

1. There is a unitary interaction that models Eve answering Adam's question about the definiteness of her *e*-spin belief.

2. Eve's answers are *always* given honestly and competently on the basis of what her beliefs are—or are not—as determined by the Bare theory's standard thinking about superpositions. So she will answer Yes if in an eigenstate of definite *e*-spin belief, and No if not.

3. The model must allow Adam (or Eve, if she is just introspecting about the structure of her own beliefs) to distinguish situations in which Eve chooses to respond (or conclude) Yes from situations in which she chooses to respond (or conclude) No.

The inconsistency follows easily. Let $|\text{'Huh?'}\rangle_{\text{Eve}}$ be any nontrivial superposition of $|\text{'Believes } e\text{-spin } up\text{'}\rangle_{\text{Eve}}$ and $|\text{'Believes } e\text{-spin } down\text{'}\rangle_{\text{Eve}}$. Assumptions 2 and 3 entail that the unitary evolution that 1 assumes to exist must, in particular, map:

$$
\begin{aligned}
&|\text{'Believes } e\text{-spin } up\text{'}\rangle_{\text{Eve}}\, |\text{'Ready to Answer'}\rangle_{\text{Eve}} \\
&\rightarrow |\text{'Believes } e\text{-spin } up\text{'}\rangle_{\text{Eve}}\, |\text{'Yes, I have a definite } e\text{-spin belief'}\rangle_{\text{Eve}}, \\
&|\text{'Huh?'}\rangle_{\text{Eve}}\, |\text{'Ready to Answer'}\rangle_{\text{Eve}} \qquad (7)\\
&\rightarrow |\text{'Huh?'}\rangle_{\text{Eve}}\, |\text{'No, I don't have a definite } e\text{-spin belief'}\rangle_{\text{Eve}},
\end{aligned}
$$

which it cannot, since that would be to map initially nonorthogonal vectors onto orthogonal ones. (The contradiction continues to hold without the idealization that in responding Eve fails to disturb her own state of belief.)

So from the point of view of the Bare theory, if we understand the theory as requiring 1, 2, and 3 above for a "good" measurement of the definiteness of Eve's *e*-spin belief, there *is no* good way to model Eve's response quantum-mechanically.

Nevertheless, since quantum mechanics is everything on the Bare theory, and we think we *do* ascertain the definiteness of each other's (and, indeed, our own) beliefs via interactions of *some* sort, those interactions must be capable of being modeled by *some* kind of unitary evolution. So 1 is not open to question. And 3 cannot be given up without undermining the possibility of Eve ascertaining and reporting on what the structure of her *e*-spin beliefs is, which is presumably something we *do* think she can do. So 2 needs to be loosened up in some way.

But in what way? Albert takes it that the way to model Eve's response unitarily is via an interaction of type (5). He appeals to the competency Eve would have to report her mental state when it is an eigenstate of some definite belief about *e*-spin. He then rigs the unitary interaction of (5) so that it correlates the answer Yes to those eigenstates,[7] concluding (by linearity) that Eve must be completely *in*competent in reporting her mental state when in a superposition of them.

However, the following parallel line of argument is equally compelling, and equally natural in the context of the Bare theory. Suppose, instead, that Eve is competent to report her mental state when it is *not* an eigenstate of some definite belief about *e*-spin. (After all, why assume a priori that she would somehow be speechless about her *lack* of definite belief?) Let $|\text{'Huh?'}\rangle_{\text{Eve}}$ and $|\text{'Wha?'}\rangle_{\text{Eve}}$, not eigenstates of definite belief about *e*-spin, span the same plane in the Hilbert space that represents Eve's brain states as the eigenstates $|\text{'Believes } e\text{-spin } up\text{'}\rangle_{\text{Eve}}$ and $|\text{'Believes } e\text{-spin } down\text{'}\rangle_{\text{Eve}}$ do. Then in each of those noneigenstates of *e*-spin belief, Eve should respond No to the question, "Do you have a definite *e*-spin belief?" By linearity, she will also respond No for any brain state in the plane. And that includes the very states in which the Bare theory guarantees that she *will* have a definite belief, namely, $|\text{'Believes } e\text{-spin } up\text{'}\rangle_{\text{Eve}}$ and $|\text{'Believes } e\text{-spin } down\text{'}\rangle_{\text{Eve}}$!

For example, if

$$
\begin{aligned}
|\text{'Huh?'}\rangle_{\text{Eve}} &= \frac{1}{\sqrt{2}}\,|\text{'Believes } e\text{-spin } up\text{'}\rangle_{\text{Eve}} + \frac{1}{\sqrt{2}}\,|\text{'Believes } e\text{-spin } down\text{'}\rangle_{\text{Eve}}, \\
|\text{'Wha?'}\rangle_{\text{Eve}} &= \frac{1}{\sqrt{2}}\,|\text{'Believes } e\text{-spin } up\text{'}\rangle_{\text{Eve}} - \frac{1}{\sqrt{2}}\,|\text{'Believes } e\text{-spin } down\text{'}\rangle_{\text{Eve}},
\end{aligned}
\tag{8}
$$

then

$$
\begin{aligned}
|\text{'Believes } e\text{-spin } up\text{'}\rangle_{\text{Eve}} &= \frac{1}{\sqrt{2}}\,|\text{'Huh?'}\rangle_{\text{Eve}} + \frac{1}{\sqrt{2}}\,|\text{'Wha?'}\rangle_{\text{Eve}}, \\
|\text{'Believes } e\text{-spin } down\text{'}\rangle_{\text{Eve}} &= \frac{1}{\sqrt{2}}\,|\text{'Huh?'}\rangle_{\text{Eve}} - \frac{1}{\sqrt{2}}\,|\text{'Wha?'}\rangle_{\text{Eve}}.
\end{aligned}
\tag{9}
$$

If Eve (competently) responds No to the question about the definiteness of her *e*-spin belief in the states $|\text{'Huh?'}\rangle_{\text{Eve}}$ and $|\text{'Wha?'}\rangle_{\text{Eve}}$, by linearity she will (*in*competently) respond No to the question in the states $|\text{'Believes } e\text{-spin } up\text{'}\rangle_{\text{Eve}}$, $|\text{'Believes } e\text{-spin } down\text{'}\rangle_{\text{Eve}}$.

So two completely parallel arguments for how to model Eve's response yield utterly incompatible results and incompatible stories about the way in which Eve is deluded. The Bare theory per se does not tell us which is the correct model, because it contains no principle that stipulates in what ways experimenters should be competent to report their beliefs, and in what ways they should not. In other words, the Bare theory contains no prescription that says that when Eve reflects about her own beliefs, Eve's brain is hard-wired so that Albert's unitary evolution (5) gets turned on inside her brain, as opposed to the unitary evolution just described. The Bare theorist cannot treat evolution (5) as a kind of calibration condition that defines what we mean by Eve's response acting as a good measurement of the structure of her own beliefs. For the evolution which induces her to respond No in both the $|\text{'Huh?'}\rangle_{\text{Eve}}$ and $|\text{'Wha?'}\rangle_{\text{Eve}}$ states is based on an equally good calibration condition, given that under the Bare theory the No response would be the correct one for those states.

However, this is still not enough to reject the Bare theorist's story out of hand, for there is an easy way Albert can grant the point just made. No interpretation of quantum mechanics should ever be called upon to decide a priori how to model a given interaction. So, in this case, we are free to find a model that fits the data that we never find people reporting that they fail to have definite beliefs about things when they are in eigenstates of definite belief—assuming that is an empirically established (or, at any rate, establishable) fact. Then, the ability of the Bare theory to at least represent (if not strictly deduce) the fact that experimenters never take their beliefs to be indefinite is perhaps all that should be required.

But before this aspect of the Bare theory is shouted from the rooftops, we would do well to put it into perspective. Compare what remains of the Bare theory's explanatory strategy to the strategy adopted by nonstandard interpretations of no-collapse quantum mechanics that inject variables, worlds, or minds into the discussion to make definite Eve's beliefs in a superposition such as (4).

The chief reason why standard thinkers about superpositions dislike the nonstandard strategy is that it requires certain observables—such as the observable whose eigenstates underpin Eve's beliefs—always to have well-defined values, even when the observable's eigenstates are superposed by the quantum state of the system. Standard thinkers do not accept that sort of discrimination between physical observables: as far as the formalism of quantum mechanics is concerned, all of its observables are on a par. To single out whatever observable it is that grounds our beliefs as an observable that it always makes sense to ascribe a definite value to regardless of quantum state is regarded as irredeemably ad hoc. There is just nothing special about that observable as compared to any other; nothing "preferred" about the set of belief eigenspaces in the Hilbert space representing Eve's brain over any other observable's eigenspaces in that space. (And, of course, if we assume all observables have simultaneously definite values all the time, that all sets of eigenspaces are on a par with respect to determinacy, then we run into well-known difficulties with what we mean by saying that every observable has a determinate value [see note 10].)

In that regard, however, the Bare theory's strategy for dealing with the problem of what experimenters take themselves to believe on the basis of their measurement devices is now no less objectionable than the nonstandard approach. The nonstandard preferred-observable approach would say that it just so happens in our world that "belief" is the kind of physical observable for which it always makes sense to speak about having one belief or another (where that may, in purely physical terms, amount to saying that the particles in experimenters' brains take up one configuration over another, if what we call beliefs turn out to be nothing other than certain positions being taken up by certain brain particles). The Bare theorist has to say that it *just so happens* in our world that the brain is hard-wired to give us the kind of beliefs we take ourselves to have about our beliefs.

In fact, the analogy between the two explanatory strategies is far closer than that. The difference between the unitary evolution that would make Eve respond Yes and the one that would make her respond No is just a difference in choice of basis: either we decide that she is in fact competent to report the definiteness of her belief in both of two orthogonal belief eigenstates, or we decide that she is in fact competent to report her lack of definite belief in both of two orthogonal *non*eigenstates of belief that span the same plane. The comparable choice on a preferred-observable approach is between Eve's belief observable as determinate versus, say, the observable with $|\text{'Huh?'}\rangle_{\text{Eve}}$ and $|\text{'Wha?'}\rangle_{\text{Eve}}$ as eigenstates. The formal structure in quantum mechanics that needs to be distinguished by the Bare theory to get a fix on modeling observers' reports *is exactly the same structure* preferred-observable approaches exploit, namely a preferred basis in the subspace spanned by $|\text{'Believes } e\text{-spin } up\text{'}\rangle_{\text{Eve}}$ and $|\text{'Believes } e\text{-spin } down\text{'}\rangle_{\text{Eve}}$ picked out by those two vectors.

So despite its name, the Bare theory cannot get by with less than nonstandard preferred-observable approaches do in explaining why experimenters report themselves to have definite beliefs. It cannot be claimed that the Bare theory's choice is determined by empirical facts any more than that can be claimed for the preferred-observable approaches. The fact that experimenters respond Yes in belief eigenstates, and also respond like that in any state that is a linear superposition of those eigenstates, is exactly the sort of empirical fact preferred-observable approaches can point to in justifying their own particular preference, namely, for the belief observable as the one that's determinate.

3. The Bare Theory Exposed

So far we have argued that if the Bare theory's story about how experimenters come to believe/report that their beliefs are definite is granted, then it is no better off than preferred-observable approaches with respect to the structure in the quantum formalism that needs to be distinguished. We now turn to arguing that the Bare theory is actually worse off because it does not succeed in explaining all that needs to be explained. The final section of the chapter will point out how the preferred-observable approach succeeds exactly where the Bare theory falls short. The upshot will be that this entire exercise concerning the Bare theory supplies a nice (even if somewhat scholastic) argument for a preferred-observable approach to universal no-collapse quantum mechanics.

It will be important, first, to be absolutely clear about what the standard thinking about quantum states is, since our contention will be that Albert does not follow that thinking to its logical conclusion.

There is no difficulty in capturing what standard thinking is about eigenstates. If a physical system is in an eigenstate of an observable, then that observable possesses the corresponding eigenvalue (or at least the system has the surefire disposition to behave as if that value were a property it possesses). In superpositions of two (or more) distinct-eigenvalue eigenstates of an observable, things are more delicate.

Suppose the eigenstates' eigenvalues are x and y. Then standard thinking entails that it would not be right to say in the superposition that the observable has value x, nor would it be right to say it has value y, nor would it be right to say it has *both* values, nor would it be right to say it has *neither* value. This is Albert's own understanding of standard thinking about superpositions (1992, 11, 79n). However mind-boggling these assertions might appear, they are the rules of the game that need to be followed if the Bare theory is to be entertained.

Now in our thought experiment about Eve, Albert makes a conceptual distinction between her brain as a repository of beliefs about *e*-spin, and the aspects of her that reflect on her beliefs about *e*-spin and then report on her reflections if she is asked questions about her beliefs. The fact that she verbalizes her beliefs about her *e*-spin beliefs is unimportant, although at one point Albert suggests it might be important: "[Eve] is necessarily going to be convinced (or at any rate she is necessarily going to *report*) that she *does* have a definite particular belief" (1992, 118). It cannot possibly be important, because if she does not *think* herself to have a definite *e*-spin belief, then that is presumably going to be a problem for the Bare theory regardless of what she reports. Furthermore, the very act of her reporting that she has a definite *e*-spin belief should presumably induce her to give assent to the content of her report, if the way we are modeling Eve as a quantum-mechanical automaton is going to have any relation to our own experience that we (in normal circumstances) don't bear witness to things we aren't prepared to accept ourselves.

So we can view the eigenstates of Eve's *reporting* Yes that we introduced earlier as eigenstates of her *reflecting* Yes (but not necessarily verbalizing the answer) without compromising the Bare theory in any way. Then what Albert needs is a conceptual distinction between Eve's brain as a retainer of beliefs about things such as *e*-spin and Eve's brain as it functions in reflecting about the structure of those beliefs. For easy reference, call these two faces of Eve the "inner" and "outer" Eve respectively.[8] And let's grant to Albert that they will interact in exactly the way he postulates, that is, in accord with equations (5).

For Albert to make his point that the Bare theory predicts that Eve will be deluded about the definiteness of her *e*-spin beliefs, he needs to establish that the outer Eve draws a conclusion at variance with the Bare facts of the matter about the inner Eve. And for both inner and outer Eve, Albert must employ standard thinking to determine the Bare facts of the matter about Eve's inner beliefs, and about her outer beliefs about her inner beliefs.

The inner facts of the matter in the belief superposition are, in particular, that Eve doesn't believe *up* and she doesn't believe *down*. These are quite specific negative claims about her situation. What we are leading up to is that Albert unjustifiably pulls back from drawing a similar kind of negative conclusion when outer Eve has the potential of getting into the same kind of superposition.

When outer Eve reflects upon whether she has exactly one of the two beliefs in the set {Believes *e*-spin *up*, Believes *e*-spin *down*}, she always concludes Yes (on the assumption that a particular unitary interaction, namely (5), is singled out as modeling this act of reflection). So if the Bare theory is true, Albert is right to observe that Eve is deluded when answering Yes in the superposition. But if the Bare theory *is* true, we can also ask what it will predict when Eve attempts to reflect upon what belief about *e*-spin she has. Since she would then get into a superposition of believing that she believes *up* and believing that she believes *down* (assuming she is good at ascertaining her specific beliefs in the eigenstate cases), under the Bare theory she will be unable to specify *which* of the two beliefs she takes herself to hold.

Recall what Albert says about that when Eve was being asked to report to Adam her reflections about what *e*-spin belief she has while in the superposition: "it won't be any easier to interpret a 'response' like that than it was to interpret the superposition of brain states in [(4)] that that response was intended to be a *description* of!" True enough. But the general difficulties of interpreting superpositions notwithstanding, the Bare theory has something quite specific to say about Eve getting into a superposition of believing that she believes *up* and believing that she believes *down*. The Bare theory says that outer Eve will *not* believe that she believes *up* and will *not* believe that she believes *down*, in just the same way that it says that inner Eve does not believe *up* and does not believe *down*.

The point is that we don't merely take ourselves to have "definite beliefs" in everyday life. After performing spin measurements on electrons, we sometimes take ourselves to believe *up* and other times take ourselves to believe *down*, depending on how our experiments go. The Bare theory can only explain *that* when we perform measurements on electrons in spin eigenstates—not in superpositions. The issue that seems to have been lost sight of in all of this is that we *also* commonly believe that we have the *grounds* for believing that we have definite beliefs, in the sense that on any given occasion we are able to reflect on what our belief is!

Let's put it another way. Earlier in his book (1992, 78–79), Albert takes the problem of interpreting a universal quantum mechanics without the collapse postulate to be that the theory predicts we should have no particular belief in a superposition when, by "direct introspection," we apparently do. His strategy for the Bare theory is to see if we can get it to explain why we take ourselves to have a definite belief in a superposition. But Albert has merely solved a watered-down version of the problem. The problem is not just to explain why we take our beliefs about things like *e*-spin to be definite in some noncommittal sense, but why we take ourselves to believe *the specific things that we do apparently believe* about *e*-spin! In trying to get the Bare theory to explain *that*, it will (as Albert himself admits) land outer Eve into the same sort of devastating superposition that inner Eve is in. But that fact is hardly irrelevant to the Bare theory's inability to explain what we take our specific beliefs to be.

The Bare theory's story about apparently definite beliefs is so seductive (as each of the present authors knows from personal experience) that anticipating responses to this quite elementary criticism we have just made is obligatory.

Note, first, that we are not accusing the Bare theory of any flat-out logical contradiction with respect to Eve's beliefs. She does believe she has exactly one of two *e*-spin beliefs, even while:

she doesn't believe that she has the *up* belief; and

she doesn't believe that she has the *down* belief. (10)

This *is* an extraordinary state of affairs, the likes of which we just don't ever seem to find ourselves in. But surely it is no more logically contradictory than asserting the provability within some formal system of "$G \vee \neg G$"—where G is a Gödel sentence of that formal system—while denying the provability of G and the provability of $\neg G$.

On the other hand, we are *not* making the move from (10) to:

she believes that she *doesn't* have the *up* belief; and

she believes that she *doesn't* have the *down* belief. (11)

The pair of conclusions about Eve in (11) obviously do logically contradict her taking herself to have exactly one of the two *e*-spin beliefs. But to draw either of them on the Bare theory when Eve is in the superposition, outer Eve would have to get into an eigenstate of belief about something particular about *e*-spin that she doesn't believe. There has been no suggestion of that possibility here.[9]

Nor do we need to establish anything like that. It suffices for our critique that the Bare theory is unable to represent the fact that we do sometimes take ourselves to believe a specified thing such as *up* when, according to the theory, we are in a state like (4). That is, it suffices to point out that the Bare theory's standard thinking leaves Eve with *no impression whatsoever* about whether or not she believes *up* on any particular occasion, even granting that she is mistakenly convinced that her *e*-spin beliefs are definite on all occasions.

Now we have granted Albert that Eve will mistakenly (but without being aware of her mistake) believe that she either believes *e*-spin to be *up* or believes it to be *down* (which is, of course, what believing that her *e*-spin belief is definite is all about). But then if she is rational and knows in advance that she is measuring a spin$-\frac{1}{2}$ particle, won't she also sometimes conclude that she believes *up*, and other times conclude that she believes *down*, even while she is in the superposition? That seems to be the obvious way to read Albert's claim that Eve "effectively knows" *e*'s spin.

But it's not on, for the simple reason that we cannot just endow Eve with magical powers of reasoning that transcend her states of belief as determined by the Bare theory. If we say that on some occasion while in the superposition Eve draws the conclusion that she believes *up* from her sense that her beliefs about *e*-spin are definite, we are saying something that flatly contradicts the Bare theory unless she can get into an eigenstate of drawing that conclusion. But the only way she can do that is if she is actually in an eigenstate of believing *up* and *not* in the superposition, as we have already emphasized.[10]

Another way to avoid our criticism might be for Albert to introduce a third level of Eve (giving us now the *three* faces of Eve) that reflects upon what her reflections about her beliefs are. When asked in the superposition if her reflection on what her *e*-spin belief is agrees with her actual *e*-spin belief, Eve can be modeled so as to respond (you guessed it) *Yes*. In fact, this is exactly what Albert uses to explain the repeatability of measurements, and the agreement between different experimenters who measure the same electron as to what its spin is (1992, 119–20).

But that's not on either. Granted the Bare theory can explain intersubjective agreement between experimenters about what their measurement outcomes are. But what cannot be explained is why sometimes the experimenters will take that agreement to be established because both their results were *up*. Mere "agreement" is not enough. The same goes for the three faces of Eve. Even if her third face is convinced that her second- and first-level selves are in accord about her particular *e*-spin belief, that does nothing to give her any sense of what those selves have agreed to. Indeed, agreement tests between first-, second-, third- . . . nth-level Eve could never be used as the factual basis for her feeling or belief that she believes *up* after some particular measurement trial, simply because in superposition (4) she can never get into any sort of eigenstate that makes reference to just the one particular belief *up*.[11]

At this point the reader might have become suspicious about our introduction of inner and outer Eve, and that our troubles with the Bare theory stem from that naive conception of Eve. All there is to Eve is her beliefs and the way she reports on them, and nothing in between! That is certainly more faithful to Albert's *text*. But it does little to mute the problem; it only changes the terms in which the problem needs to be stated. The problem would then become that, even if Eve operationally qualifies as a person convinced of the definiteness of her *e*-spin belief, in the superposition she will never be able to qualify as a person convinced she has the particular belief that *e*-spin is *up*. Talking about an introspective side of Eve versus what an external observer would infer about what she takes her mental state to be changes nothing. And this is so even if we demand—as Albert's phrasing of Adam's question does—that Eve give verbal assent to the fact that her *e*-spin belief is "not uncertain and not confused and not vague and not superposed."

So much for anticipating possible responses to our main criticism. The same criticism undermines the story Albert runs to explain why, on the Bare theory, experimenters take themselves to have confirmed the statistical predictions of quantum mechanics through their measurements (1992, 120–23). In the limit of an infinite sequence of measurements, an experimenter's brain will get into a quantum state that lies in an eigenspace of the relative frequency operator, with an eigenvalue equal to the quantum-mechanically predicted frequency for the measured observable. And so, according to the Bare theory, experimenters will take themselves to have obtained sequences that agree, in their long-run frequencies, with quantum statistical predictions.

But experimenters as we know them seem to believe far more than that. They take themselves to have obtained particular sequences with particular characteristics. Merely getting into an eigenspace of the relative frequency operator will not ensure *that* on the Bare theory, since there are numerous distinct sequences that can yield the same long-run frequencies. Albert's own problem with this story is that it is too idealized. Determinateness with respect to the relative frequency operator is only achieved in the limit of an infinite number of measurements (1992, 124–25). But even the Bare theory's idealized experimenter is unlike any we are ever likely to meet (or be).

A similar point is made by Stein (1984) against Geroch's (1984) reading of Everett's interpretation (which is essentially the Bare theory; see notes 2 and 3). Stein asks how we ordinarily go about checking that measurement outcomes conform to the relative frequencies predicted by quantum mechanics, and answers (1984, 641, emphasis in the original): "We do so by performing the experiments and *noting and counting their outcomes*." We too have argued that *that* part of our ordinary experience cannot be accounted for on the Bare theory. However, Stein goes further than he should (or needs to) in suggesting that Geroch's observers are committed to the exclusive use of physically special apparatus designed to record only statistical information and "obliviate" particular outcomes:

> Geroch would have us (conceptually at least) interpose, between the experimenters and the *real* observation, apparatus that is designed just to detect whether or not the predicted frequencies (within the predicted range of variability) have occurred: what we may call "obliviating apparatus," designed expressly to ignore everything about the series of experiments that was not predicted or precluded with practical certainty by quantum mechanics. The idea that physicists might actually proceed in this way—spend money and time on instrumentation (automatic recording apparatus to feed data to computers, programs designed to report only the predicted aspects of the data and to destroy all records of details that were not predicted)—simply in the interest of conformity to the Everett-Geroch interpretation, seems worthy of Swift's Academy of Lagado. (1984, 642)

In spite of the initial qualification, it appears that Stein regards the obliviating apparatuses required by the Bare theory (in Albert's or Geroch's version) as more than merely conceptual. This goes too far, since the Bare theory has no need to postulate devious devices designed specifically to lose or erase information about outcomes while retaining statistical information. The fact that an apparatus, or an observer's mind, only retains long-run statistical information follows quite straightforwardly from the observation that information is registered only if the state of the registering system is an eigenstate of registering it! This is a direct consequence of the standard thinking about superpositions employed in the Bare theory and needs no further special physical assumptions about the design of measurements, beyond those commonly granted in standard analyses of quantum measurement. So while we agree with Stein's starting point, we believe he overstates the case.

Note, finally, that an idealized experimenter measuring an ensemble of electrons in superposition (2) will, in the limit, approach an eigenstate of answering Yes to the question, "Did you sometimes take yourself to believe *up* on the basis of your measurement result?" So isn't it the case, after all, that the Bare theory *can explain* why we sometimes take ourselves to believe *up* (and other times *down*) on the basis of our measurements? No. The alleged explanation here trades on an ambiguity in the word "sometimes." What's important is not just whether at some point or other (sometimes) in the experimenter's measurement outcome sequence an *up* was taken to be registered, but whether on (say) trial number 13 (some particular time) the experimenter took him- or herself to have measured *up*. Again, getting into an eigenspace of the relative frequency operator will not suffice for a definite answer to *that* question.

We conclude that the Bare theory fails to explain why we take ourselves to have specific beliefs about measurement outcomes or outcome sequences in generic experimental situations. The only conclusion left is that if the Bare theory is true, then in states like (4) we really *never* believe that we believe *up* nor do we ever believe that we believe *down* after (any single!) measurement of an electron's spin. The same goes for believing that we believe anything particular about measurement outcome sequences beyond what their long-run frequencies are (and perhaps other features of outcome sequences definable in the long run, such as their randomness). If that's "What It Feels Like to Be in a Superposition" (Albert 1992, 112), then the Bare theory bears far less on the actual experiences of actual experimenters than we have been led by Albert to believe.

4. A Fig Leaf for the Bare Theory

Fortunately, explanatory adequacy with regard to what we take ourselves to believe is quickly restored on a preferred-observable approach to universal, no-collapse quantum mechanics that simply denies standard thinking about superpositions from the very beginning.[12]

On such an approach, all the endless problems about beliefs about beliefs and responses about beliefs engendered above are cut off at their roots. Because any observable representing beliefs is granted the status of an observable that is always determinate, in superposition (4) inner Eve will believe $|c_1|^2$ of the time that *e*-spin is *up* and $|c_2|^2$ of the time that *e*-spin is *down*. Outer Eve will believe that she has definite beliefs, in agreement with the Bare theory. But she will also take herself to believe *up* $|c_1|^2$ of the time and to believe *down* $|c_2|^2$ of the time. And this is exactly in accord with *all* the things experimenters take themselves to believe. They will not be left (unknowingly) deluded, but more important, not be left without a sense of what they take their specific beliefs about *e*-spin to be.

So making belief the preferred observable (whatever the physical basis of our beliefs turns out to be) is exactly the fig leaf the Bare theory needs to fill the gap between guaranteeing that experimenters have the sense that their beliefs are definite, and giving them the specific feelings they get about the specific beliefs they hold.

Albert's Bare theory argument concerning what Eve takes herself to believe, the argument that we have spent the bulk of this paper analyzing, goes back at least as far as the paper by Albert and Loewer (1988). Interestingly, they do not treat the Bare theory as a possible end in itself, but use it as a springboard to draw a conclusion that is the same in spirit as our own.

Albert and Loewer use the Eve scenario to argue that if we maintain the idea that experimenters never report that their beliefs are definite when that is in fact false, then their belief states cannot supervene on a superposition like (4), but must be specified in addition to it (1988, 205). In other words, accurate reporting requires a preferred-observable approach with the observable that underpins Eve's beliefs always taken to be determinate independently of the quantum state of Eve.[13]

We agree with Albert and Loewer's conclusion that the Eve scenario most naturally points to a preferred-observable approach, and that assuming reports of belief definiteness are accurate (and there is no collapse) entails that conclusion. But we have shown that one can arrive at that conclusion without any such assumption. Our point has been that the Bare theory is too bare to account for the specificity of what we take ourselves to believe, and that *actual* determinateness of beliefs (as opposed to *effective* determinateness) is necessary to account for that specificity.

The Bare theory's Adam and Eve fall short of what we take ourselves to be like. Having eaten of the fruit from the tree of the knowledge of *up* and *down*, we know what's up.

Notes

We would like to thank John Bell, Jeff Barrett, Geoffrey Hellman, Tim Kenyon, and Steve Weinstein for helpful discussions (without committing them to endorsing anything that has been argued here). R. C. acknowledges support from the Vice President of Research at the University of Western Ontario, and B. M. acknowledges a National Science Foundation Graduate Research Fellowship.

1. Writing down a ket like $|$'Believes *e*-spin *up*'$\rangle_{\text{Eve}}$ carries with it no commitment to there being a unique way to instantiate a state of Eve physically that would count (by operationalist or other criteria) as a state of her believing *up*. Furthermore, the quantum-mechanical observable that underpins "belief," whatever it is, is bound to be degenerate simply on the grounds that any particular belief, like the belief that *e*-spin is *up*, may be held concurrently with a host of other beliefs about other topics—beliefs that will vary from one experimenter to another.

2. Albert (1992, 124) sees the Bare theory as an attractive way to elaborate Everett's (1957) ideas as a "one world" theory that preserves the standard way of thinking about superpositions (that is, a theory based on the "eigenvalue-eigenstate link" as the criterion for value-definiteness or determinateness). While Albert cites Lockwood (1989), who follows Deutsch (1985), Geroch (1984) has also independently proposed an interpretation of Everett that is very much along the same lines. For more recent technical and conceptual

elaborations of this sort of "bare" interpretation of Everett, see Barrett (1994, 1995, 1996). Note that these interpretations differ from Bell's (1987) "one world" interpretation of Everett. Bell's interpretation—as he puts it (1987, 133), "the pilot-wave theory without the trajectories"—is a preferred-observable interpretation of the type considered in section 2, with position in configuration space as the preferred, always determinate observable. We shall not be concerned here with the murky topic of Everett exegesis.

3. Geroch (cf. previous note) puts the aim of this explanatory project well: "what must be accounted for, if anything, is, not the specific classical outcomes deemed to have occurred for a specific experiment, but rather the general human impression that classical outcomes do occur" (1984, 628–29). But, unlike Albert, Geroch does no more than sketch how one might try to complete the project, and remains agnostic about its feasibility. Thus he continues: "This problem may well be soluble, but is probably beyond our present abilities; and, in any case, is basically not a problem in quantum mechanics" (1984, 629).

4. It is not important for what follows that the particulars of *how* Eve responds be the same in the two eigenstates of her *e*-spin belief (that is, responding with a whisper or a scream, via one set of neurons firing in her brain over another set that could equally well do the job, and so on). It is only important that the *content* of her response be the same. In other words, if we consider the subspace in the Hilbert space of Eve spanned by all those states of her that we are willing to count (by operational or other criteria) as states in which the content of her response is Yes, then what's important is that the two occurrences of the ket $|\text{'Yes, I have a definite } e\text{-spin belief'}\rangle_{\text{Eve}}$ in (5) both be vectors that lie in that subspace. They need not generate the same ray, even though we have not taken the trouble to reflect that possibility in our notation.

5. For present purposes, we can think of Eve's physical state as an element of the tensor product of the Hilbert space that represents the features of her brain that register beliefs about *e*-spin with the Hilbert space associated with the parts of Eve involved in reflecting on what her beliefs are and responding to questions about them. It does not matter whether the part(s) of Eve's brain associated with the relevant beliefs are individuated neuroanatomically or functionally, that is, whether a Token or Type physicalism is taken to apply. And our case against the Bare theory is unaffected even if one eschews mind-brain identity in favor of the neurological causes of mere belief *reports*.

6. It should perhaps be pointed out that Albert is not entirely uncritical of these stories himself (1992, 124–25). But once our "decisive" problem is laid out, it will become clear that we believe Albert should have been critical of these stories at a far earlier stage in his exposition.

7. It might have occurred to the reader that interaction (5) physically amounts to nothing more than a measurement of the value of the projection operator onto the subspace spanned by $|\text{'Believes } e\text{-spin } up\text{'}\rangle_{\text{Eve}}$ and $|\text{'Believes } e\text{-spin } down\text{'}\rangle_{\text{Eve}}$ and, as such, carries little information about Eve's physical state beyond the fact that it lies in that subspace. But Albert is well aware of this (see his particle in a box analogy [1992, 117n]). Albert's point is that Eve in the superposition will be responding to Adam's question in exactly the same way as she would if she were in an eigenstate of *e*-spin belief (at least with regard to the content of her answer; see note 4), and therefore she (and others) will be given all the same impressions about her state of mind in the two cases. What purely physical meaning one assigns to all of this, in terms of what observable of Eve has actually been measured via (5), would presumably be regarded by Albert as beside the point.

8. This really is for ease of reference: we are not suggesting that Eve has multiple personalities, but only that she is capable of forming beliefs and also (second-order) beliefs about her beliefs. That much is necessary to get the story about Eve's delusion about her beliefs off the ground.

9. Evidently, the closest one can get to (11) on the Bare theory is to have Adam ask Eve the question, "Do you *lack* exactly one of the two beliefs in the set {Believes *up*, Believes *down*}?" Eve will (presumably) respond Yes to this in the superposition if that was her response to the original question about *having* exactly one of those two beliefs (that is, setting aside the previous section's worries about Yes versus No). But clearly answering Yes to both questions is logically consistent. Nevertheless, the Bare theory again remains explanatorily inadequate with respect to the specific things about *e*-spin we take ourselves *not* to believe subsequent to a particular measurement trial on an electron's spin.

10. The fact that superposed Eve is never able to take herself to believe *up* or take herself to believe *down*, even though she takes herself to believe exactly one of those, is analogous to a well-known difficulty faced by realistic quantum logic (as defended, for example, by Putnam [1975]) in its attempt to render meaningful statements such as "Every observable has a definite value." Recall that the difficulty is that, even though the proposition "Observable O has a definite value" may be identified with the disjunction "o_1 or o_2 or . . . or o_n"—where (for any i) o_i is the proposition "Observable O has value o_i"—the truth of this disjunction, which is a tautology in quantum logic, turns out to be compatible with the falsity of all of its disjuncts. What makes this a difficulty for realistic quantum logic is that it waters down the very realism of definite values that logic was designed to prop up. By analogy, what we have argued against the Bare theory is that Eve's impression of the definiteness of her *e*-spin belief, expressed through her belief in a disjunction over her believing *up* versus *down*, necessarily yields only a watered-down representation of our actual experience, because Eve's belief cannot be accompanied, on any given occasion, by belief that a particular one of those disjuncts is true. However, there is a downside to this quantum logic analogy: we do not (and need not) claim that the Bare theory forces Eve (or Adam) to understand her answer to Adam as a conscious endorsement of quantum over classical logic. Indeed, just as Adam can ask Eve what she thinks about the definiteness of her *e*-spin belief and will always elicit a Yes response, he will always elicit that same response by asking her whether she takes the structure of her beliefs to conform to the logic of the classical "or"!

11. Jeff Barrett (in correspondence with the authors) has suggested another possible way to bring the Bare theory closer to endowing Eve with the (illusory) sense that her *e*-spin belief has the content that it should have. Let Eve make spin measurements on three identical electrons: the first in an *up* eigenstate, the second in a *down* eigenstate, and the third in a superposition thereof. Again, Eve will believe she formed a definite belief about all three measurement results. But, furthermore, she will believe that the result of her last measurement was indistinguishable (in its content and the impression it left on her) from the result of exactly one of her first two measurements! Alas, this new "agreement argument" is still too weak to explain why it is that we think we can specify which of the first two measurement results the third agreed with. Put another way, the Bare theory cannot vindicate the belief that the third result was indistinguishable from the first result, or the belief that it was indistinguishable from the second. Beliefs of this sort may well be mistakenly acquired in our world, but the Bare theory is still at a loss to explain why.

12. Note that preferred-observable interpretations, as we understand them, include the modal interpretations discussed by several contributors to this volume. In a modal interpretation formulated along the lines of the versions originally proposed by Kochen (1985) and by Dieks (1989), the preferred observable is derived from the quantum state and is therefore time-dependent, as opposed to a preferred-observable interpretation with a fixed, always determinate, observable. For a detailed analysis comprehending both these ways of choosing a preferred observable, see Bub and Clifton (1996). The situation is more complicated in Healey's (1989) modal interpretation; see Clifton (1996) for further discus-

sion. Bacciagaluppi, Elby, and Hemmo (1996) have shown how the determinateness of observables representing beliefs in the Kochen-Dieks modal interpretation requires decoherence arguments.

13. Albert and Loewer go on to argue that the best preferred-observable approach involves invoking "many-minds," but we do not wish to enter into that discussion here. Suffice it to say that we believe everything presently under discussion is neutral with respect to how one makes the preferred observable preferred, whether it be in terms of (one or many) hidden variables, worlds, or minds.

References

Albert, D. 1992. *Quantum Mechanics and Experience*. Cambridge, Mass.: Harvard University Press.

Albert, D., and B. Loewer. 1988. "Interpreting the Many Worlds Interpretation." *Synthese* 77: 195–213.

Bacciagaluppi, G., A. Elby, and M. Hemmo. 1996. "Observers' Beliefs in the Modal Interpretation of Quantum Theory," manuscript.

Barrett, J. 1994. "The Suggestive Properties of Quantum Mechanics without the Collapse Postulate." *Erkenntnis* 41: 233–252.

———. 1995. "The Single-Mind and Many-Minds Versions of Quantum Mechanics." *Erkenntnis* 42: 89–105.

———. 1996. "Empirical Adequacy and the Availability of Reliable Records in Quantum Mechanics." *Philosophy of Science* 63: 49–64.

Bell, J. S. 1987. *Speakable and Unspeakable in Quantum Mechanics*. Cambridge: Cambridge University Press.

Bub, J., and R. Clifton. 1996. "A Uniqueness Theorem for Interpretations of Quantum Mechanics." *Studies in the History and Philosophy of Modern Physics* 27: 181–219.

Clifton, R. 1996. "The Properties of Modal Interpretations of Quantum Mechanics." *British Journal for the Philosophy of Science* 47: 371–98.

Deutsch, D. 1985. "Quantum Theory as a Universal Physical Theory." *International Journal of Theoretical Physics* 24: 1–41.

Dieks, D. 1989. "Quantum Mechanics without the Projection Postulate and Its Realistic Interpretation." *Foundations of Physics* 19: 1397–1423.

Everett, H. 1957. "Relative State Formulation of Quantum Mechanics." *Reviews of Modern Physics* 29: 454–62.

Geroch, R. 1984. "The Everett Interpretation." *Noûs* 18: 617–33.

Healey, R. 1989. *The Philosophy of Quantum Mechanics*. Cambridge: Cambridge University Press.

Kochen, S. 1985. "A New Interpretation of Quantum Mechanics." In P. Lahti and P. Mittelstaedt, eds., *Symposium on the Foundations of Modern Physics*. Singapore: World Scientific, 151–69.

Lockwood, M. 1989. *Mind, Brain, and the Quantum*. Oxford: Oxford University Press.

Putnam, H. 1975. "The Logic of Quantum Mechanics." *Philosophical Papers vol. 1*. Cambridge: Cambridge University Press, 174–97.

Stein, H. 1984. "The Everett Interpretation of Quantum Mechanics: Many Worlds or None?" *Noûs* 18: 635–52.

Richard A. Healey

"Modal" Interpretations, Decoherence, and the Quantum Measurement Problem

1. Introduction

There is a serious and unresolved quantum measurement problem. Some, like Ghirardi, Rimini, and Weber (1986), try to solve it by modifying quantum mechanics. If successful, such attempts would result in a theory, distinct from but closely related to quantum mechanics, that is no longer subject to a measurement problem. But the conceptual problem—that of reconciling quantum mechanics itself with the possibility of the very measurements we take to warrant its acceptance—would remain. That problem may be insolvable. But it is my belief that recent developments, some formal, others conceptual, promise hope for a solution.

So-called "modal" interpretations of quantum mechanics[1] offer the prospect of more determinateness in the values of observables than would be consistent with a more orthodox view of the quantum state, without introducing any new variables and without conflict with no-hidden-variables results. The hope is that the adoption of such an interpretation would make it possible consistently to maintain that a quantum-mechanically described apparatus always records a definite outcome, thereby explaining why quantum measurements are never observed to yield indeterminate results.

Now if a quantum measurement is to yield an observable result, then it seems that at some stage that result must be recorded in a directly observable feature of a macroscopic object—a "pointer position," in the familiar paradigm. And decoherence theorists have shown in a number of models that interactions with its environment typically ensure that the state of a macroscopic object will deviate at most for an extremely short time from a mixture that is very nearly diagonal in a particular basis—the so-called "pointer basis."[2]

Although it is tempting to conclude that decoherence thereby solves the measurement problem, I follow others[3] in arguing that that conclusion is mistaken. On the other hand, a number of criticisms of "modal" interpretations[4] have cast serious doubt on the claim that a solution to the measurement problem emerges automatically within such an interpretation. The arguments leading to these negative conclusions are reviewed in sections 2 through 5 of this chapter.

More recent work[5] has examined the effects of decoherence in the context of a "modal" interpretation, with the intention of rebutting objections to earlier attempts to use such an interpretation to solve the measurement problem. This work is reviewed in section 6, which applies decoherence within a "modal" interpretation to a simple model, to illustrate what I believe currently is the most promising approach to the quantum measurement problem. Section 7 addresses an important residual problem in relating the supposed definiteness of the "pointer position" to the observer's definite perceptual experience of that "pointer position."

2. The Quantum Measurement Problem

The quantum measurement problem has several aspects. The first is a semantic problem: to specify precisely what is meant by terms such as 'measurement,' as these feature in the probabilistic predictions of quantum mechanics. The second is a consistency problem: to show how the occurrence of purely physical interactions suited to serve as quantum measurements is consistent with quantum mechanics, including its linear dynamics. A third aspect of the measurement problem is to show that quantum mechanics can consistently model *actual* quantum measurements, of the kind whose occurrence we take to provide the statistics that confirm the probabilistic predictions of quantum mechanics.

There is also a closely related problem that I shall call the *determinateness problem*. Albert and Loewer (1993) characterize this problem as follows:

> [The] problem . . . is *not* merely to account for the determinateness of the positions of pointers at the conclusions of ideal non-disturbing measurement-interactions, but to account for the determinateness of whatever actually happens to be determinate, under whatever *circumstances* that determinateness actually happens to *arise*. Manifestly, no interpretation of quantum theory which offers anything less than an account of all that (whatever the problem of producing such an account is called) will do.

The determinateness problem is both wider and less well defined than the quantum measurement problem. It is less well defined because it is far from clear what is determinate, and in what circumstances. For example, it is standardly assumed that gross macroscopic objects always have quite determinate positions. But there are interpretations that deny that this is so, while seeking nevertheless to explain our experiences of macroscopic objects.

The measurement problem, in its third aspect, becomes a relatively well-defined subcase of the determinateness problem if one makes the standard assumption that at the conclusion of an actual quantum measurement the "pointer" does have a determinate "position." Moreover, no interpretation of quantum mechanics will do unless it can either solve or dissolve this subcase of the determinateness problem. An adequate interpretation must account for the fact that quantum mechanics is an empirically adequate theory, and its probabilistic predictions about "pointer positions" at the conclusion of actual quantum measurements are the core of the theory's empirical content.

Combining the semantic and consistency aspects, one arrives at the following formulation of the quantum measurement problem: to say exactly what constitutes a measurement of a quantum-mechanical observable, and then to explain how it is possible for a properly conducted quantum-mechanical measurement always to yield some definite outcome. This is a problem, because plausible attempts to say in purely quantum-mechanical terms what constitutes a measurement interaction apparently imply that properly conducted quantum measurements do not always yield some definite outcome. The classic demonstration of this result is that of Wigner (1963). I begin by reviewing the argument.

On the simplest model, a quantum measurement corresponds to a quantum-mechanical interaction between an "object" system σ and an "apparatus" system α (initially in the "ready-to-measure" state χ_0^α). The interaction Hamiltonian is such as to produce the following (unitary) transformation in their joint quantum state over the course of the interaction, for each of an orthonormal basis of eigenstates χ_i^σ of the measured observable

$$\chi_i^\sigma \otimes \chi_0^\alpha \Rightarrow \chi_i^\sigma \otimes \chi_i^\alpha, \tag{1}$$

It then follows from the linearity of the Schrödinger equation that an interaction with an initially superposed state of the "object" system will proceed as follows:

$$\left[\sum_i c_i \chi_i^\sigma\right] \otimes \chi_0^\alpha \Rightarrow \sum_i c_i [\chi_i^\sigma \otimes \chi_i^\alpha], \ \sum_i |c_i|^2 = 1. \tag{2}$$

Suppose that one assumes that the quantum state vector gives a complete description of the properties of systems that it represents. More precisely, assume that dynamical variable A has a sharp (real) value a on a system just in case the quantum state of that system may be represented by a state vector that is an eigenvector of the operator $\hat{A}$ corresponding to A (this assumption has been called the eigenvalue-eigenstate link). Under this assumption, it seems clear why equation (1) would be a good candidate for an interaction suitable for measuring the value of an observable on σ with eigenstates $\{\chi_i^\sigma\}$. For if, prior to the interaction, σ had value a_i for quantity A, then the left-hand side of (1) would appropriately represent this fact, and the right-hand side of (1) would represent the fact that A still has value a_i after the interaction, while the pointer reading quantity R on α has now acquired a sharp value r_i, the eigenvalue corresponding to eigenvector χ_i^α of the associated operator $\hat{R}$. Thus this interaction, applied to a state of σ in which A has a sharp value, will leave that value undisturbed and put α, the "apparatus" system, into a state in which its pointer reading quantity has a sharp, correlated value. These desirable consequences are taken to justify treating an interaction with the properties of (1) as a model of a quantum-mechanical measurement.

But when this same interaction is applied to a state of σ that is not an eigenstate of $\hat{A}$ (as in equation [2]), the consequences of the model are much less desirable. For the eigenvalue-eigenstate link now implies not only that A has no sharp value before the interaction, but more worryingly that A has no sharp value even after the interaction, and, even worse, that the pointer reading quantity R on α has no sharp value after the interaction.

Of course, equations (1) and (2) provide a highly oversimplified quantum-mechanical model of any actual quantum measurement. A more realistic model would acknowledge the extreme degeneracy of each "pointer reading" state of an apparatus system, and would express our ignorance of its exact microstate by representing its state as a mixture over such degenerate microstates. It would further acknowledge that a quantum measurement, unlike equation (1), may well disturb the value of the measured quantity while recording it (Pauli classified this as a "measurement of the second kind"), and that most, if not all, actual quantum measurements are error-prone: there is always some chance that the pointer reading quantity R will adopt value r_i, even when the initial value of A on the object system was a_j, with $j \neq i$. But it is well known[6] that generalizing this highly oversimplified model so as to take account of these complications in no way alters the essential conclusion: as long as one assumes the eigenvalue-eigenstate link, a quantum-mechanically described measurement interaction between an apparatus system and an object system that is initially in any of a wide class of states will lead to a final state in which the apparatus's "pointer reading" quantity fails to have any sharp value.

3. Decoherence: I

The model of quantum measurement represented by equations (1) and (2) is oversimplified in one further, important respect. Any actual quantum measurement involves, at some stage, an irreversible change that produces a stable record in some macroscopic apparatus, a fact not yet represented in (1) and (2). Now it is not at all clear how quantum mechanics should, or even can, represent this fact. For example, attempts to model irreversibility as an increase in the statistical entropy of the apparatus threaten to conflict with the fact that the entropy of a pure total state must remain constant under Schrödinger evolution; and the difficulty of giving any precise and relevant quantum-mechanical explication of the term 'macroscopic' is by now notorious, given our present understanding of macroscopic quantum phenomena such as superconductivity and superfluidity. But one natural way to proceed here is to acknowledge that any device capable of incorporating a stable, observable record of a quantum measurement will, in practice, constitute an open system that is strongly coupled to its environment.

Decoherence theorists have shown in the context of simplified models that interactions with the environment can rapidly destroy interference between different "pointer position" eigenstates following a quantum measurement interaction. This has given rise to the hope, if not belief, that appeal to such environmentally induced decoherence is all that is required to solve the quantum measurement problem. I believe such hope is misplaced.

Consider, for example, the simple model discussed by Zurek (1982). Here a dichotomic observable A (such as the z-spin of a spin$-\frac{1}{2}$ particle) is measured on σ by a model apparatus α, whose pointer reading quantity R is similarly dichotomic, with values $r_{\pm}$. The effect of the interaction between σ and α is as follows:

$$\chi^{\sigma}_{\pm} \otimes \chi^{\alpha}_{0} \Rightarrow \chi^{\sigma}_{\pm} \otimes \chi^{\alpha}_{\pm} \tag{3}$$

The analog of (2) is then

$$\begin{aligned}&[c_{+}\chi^{\sigma}_{+} + c_{-}\chi^{\sigma}_{-}] \otimes \chi^{\alpha}_{0} \\ &\quad \Rightarrow c_{+}[\chi^{\sigma}_{+} \otimes \chi^{\alpha}_{+}] + c_{-}[\chi^{\sigma}_{-} \otimes \chi^{\alpha}_{-}],\ |c_{+}|^{2} + |c_{-}|^{2} = 1\end{aligned} \tag{4}$$

But now suppose that immediately after this interaction α begins to interact with its environment ϵ. Specifically, let ϵ consist of a large number N of two-state systems, the kth of which has a Hilbert space spanned by the basis $\{\kappa_{+}, \kappa_{-}\}$, and let the interaction between α and ϵ be governed by a total Hamiltonian such that, t seconds after it starts, the state of $\sigma \oplus \alpha \oplus \epsilon$ is

$$\Psi(t) = c_+\chi_+ \Pi_k \otimes [c_+^k e^{ig_kt}\kappa_+ + c_-^k e^{-ig_kt}\kappa_-] + c_-\chi_- \Pi_k \otimes [c_+^k e^{-ig_kt}\kappa_+ + c_-^k e^{ig_kt}\kappa_-] \quad (5)$$

where

$$\chi_\pm = \chi_\pm^\sigma \otimes \chi_\pm^\alpha \quad (6)$$

Then the state of $\sigma \otimes \alpha$ is represented by the reduced density operator

$$\rho = |c_+|^2 P_{\chi_+} + |c_-|^2 P_{\chi_-} + z(t)\ c_+c_-^*\ Q_-^+ + z^*(t)\ c_-^*c_-Q_+^- \quad (7)$$

where

$$z(t) = \Pi_{k=1}^N [\cos 2g_kt + i(|c_+^k|^2 - |c_-^k|^2)\sin 2g_kt], \quad Q_-^+\chi_- = \chi_+, \quad Q_+^-\chi_+ = \chi_- \quad (8)$$

Provided that the coupling constants g_k are random real-valued functions of k, and N is sufficiently large, $z(t)$ will become very small even for small values of t, and will remain small for a long time thereafter. It follows that, very shortly after the start of the interaction between α and ϵ, the density operator representing the state of $\sigma \oplus \alpha$ is, and for a long time remains, very nearly diagonal in the $\chi_\pm$ basis (although as long as N remains finite, $|z(t)|$ will return arbitrarily close to 1 at certain [much later!] times). In this sense, a consequence of the postulated environmental interaction is that the quantum-mechanical state of $\sigma \oplus \alpha$ rapidly becomes indistinguishable from a mixture of the pure states $\chi_\pm$. It is then very tempting to offer a so-called ignorance interpretation of this mixture, and so to conclude that, after a very short time, the state of $\sigma \oplus \alpha$ is almost certain to be representable by either χ_+ or χ_-, with probabilities $|c_+|^2$, $|c_-|^2$ respectively. It would then follow that the state of α is almost certain to be representable by either χ_+^α or χ_-^α, and hence (by the eigenvalue-eigenstate link) that the value of R is almost certainly either r_+ or r_-, again with approximate probabilities $|c_+|^2$, $|c_-|^2$ respectively.

Does this constitute a solution to the measurement problem, at least in the context of this simple model? As D'Espagnat (1990) argues, it does not. In his terminology, the state ρ represents an improper mixture, which consequently cannot legitimately be given an ignorance interpretation. One way to show this is to note that there exists an "observable" defined on the total system $\sigma \oplus \alpha \oplus \epsilon$ that commutes with the total Hamiltonian (including the apparatus-environment interaction) and whose (constant) expectation value in state $\Psi(t)$ differs arbitrarily far from the value calculated on the assumption that the state of α is either χ_+^α or χ_-^α, or even any pure state near (in the Hilbert space norm) to χ_+^α or χ_-^α. Of course, there may be practical, or even principled, reasons as to why such an "observable" is not in fact observable, so this discrepancy in expectation values could never in fact show up in measurements on the total system $\sigma \oplus \alpha \oplus \epsilon$. But there is a stronger objection to the proposed "solution." As Elby (1994) points out, when made explicit, it leads to a contradiction.

Note that the above argument relied on the eigenvalue-eigenstate link. This reliance was overt in the inference from the state of α's being representable by either χ_+^α or χ_-^α to the value of R's being either r_+ or r_-. This inference actually only requires the eigenstate-to-eigenvalue direction of the link. However, the measurement problem only arises in the first place if one assumes that the link also holds in the other direction. That problem, recall, was first encountered in equation (2), when the superposition on the right-hand side was taken to be inconsistent with the attribution to α of any sharp value for the pointer reading quantity R after the interaction. But why should there even appear to be any inconsistency here unless one assumes that an observable has a value on a system only if the quantum state of that system may be represented by a corresponding eigenstate of that observable, an assumption that was indeed made in the discussion following equation (2)? Hence the argument assumes the eigenvalue-eigenstate link in order to derive the conclusion that after a very short time, the value of R on α is almost certain to be either r_+ or r_-, hence sharp. But when applied directly to the state $\Psi(t)$ of $\sigma \oplus \alpha \oplus \epsilon$, the eigenvalue-eigenstate link implies that α has no sharp value for the pointer reading quantity R at any time after the interaction.[7]

4. Decoherence: II

The application of quantum mechanics to simple models to demonstrate decoherence as illustrated in the previous section is not in itself controversial. Contradictions arise only with the addition of interpretative postulates such as the eigenvalue-eigenstate link. Recently, theorists such as Zurek (1991; 1993a,b) have sought to use similar demonstrations to resolve the measurement problem in a way that is free from such contradictions. Zurek has called the result the "existential interpretation" of quantum mechanics.

The basic idea is to apply decoherence in the context of an Everett-style relative-state interpretation of quantum mechanics. Recall that, if the state of a compound system $\sigma \oplus \alpha$ is given by the right-hand side of equation (2), then the relative state of σ, relative to state χ_i^α of α, is χ_i^σ. More generally, each vector from an arbitrary complete orthonormal basis for the Hilbert space of α defines a relative state of α, to which corresponds a unique (up to phase) relative state of σ. The effect of environmentally induced decoherence is then to define a preferred basis containing (relative) "observer-memory-states" in certain open physical systems. These systems constitute the physical locus of observers: following a quantum measurement interaction, each such relative state (or at least, each such state with nonnegligible amplitude in an expansion of the universal quantum state) is, and remains, an eigenstate of some recording quantity. Applying the eigenvalue-eigenstate link to these relative states, the value of the recording quantity constitutes a stable record of what an observer associated with that state takes to be the result of the measurement. Moreover, that record is both faithful—provided that the corresponding relative state of the rest of the universe is an eigenstate of the measured observable (with eigenvalue corresponding to that value of the recording quantity of which the relative "observer-memory-state" is an eigenstate)—and intersubjectively agreed, insofar as the corresponding relative state of the rest of the universe is also an eigenstate of the recording quantity of every other "observer system" in the universe (with eigenvalue recording the very same result that is recorded in the first relative "observer-memory-state").

This idea may be clarified by reference to the example of decoherence presented in the previous section. In that example, the initial interaction between systems σ and α was accompanied by an interaction between α and its environment ϵ, modeled in a particular way. Assume, for the moment, that the quantity $z(t)$ of equation (8) is and remains zero after some time T. Then the relative state of $\sigma \oplus \epsilon$, relative to state χ^{α}_{+} of α, is given by

$$\chi^{\sigma}_{+} \Pi_k \otimes [c^k_+ e^{ig_k t} \kappa_+ + c^k_- e^{-ig_k t} \kappa_-]. \tag{9}$$

Applying the eigenvalue-eigenstate link to these relative states of $\sigma \oplus \epsilon$ and α, it follows that, relative to an "observer-memory-state" of α that contains the record r_+ for the measured value of A, the postmeasurement value of A is in fact a_+. Similarly, relative to an "observer-memory-state" of α that contains the record r_- for the measured value of A, the postmeasurement value of A is in fact a_-. This is the sense in which the record of the measurement of A is faithful.

Now suppose that a second system β interacts with σ after T, and that this interaction satisfies

$$\chi^{\sigma}_{\pm} \otimes \chi^{\beta}_{0} \Rightarrow \chi^{\sigma}_{\pm} \otimes \chi^{\beta}_{\pm}. \tag{10}$$

Suppose that β also interacts with its environment ϵ', via an interaction that can be modeled in a way exactly parallel to the interaction between α and its environment ϵ, so that after a certain later time T_1, the state of $\sigma \oplus \alpha \oplus \beta$ can be represented by the diagonal density operator

$$\rho' = |c_+|^2 P_{\chi'_+} + |c_-|^2 P_{\chi'_-} \tag{11}$$

where

$$\chi'_{\pm} = \chi^{\sigma}_{\pm} \otimes \chi^{\alpha}_{\pm} \otimes \chi^{\beta}_{\pm}. \tag{12}$$

Then, relative to the state of α after T_1 that contains the record $r_+(r_-)$ for the value of A measured by α, the relative state of $\sigma \oplus \beta \oplus \epsilon \oplus \epsilon'$ will, via the eigenvalue-eigenstate link, make it true that the "observer-memory-state" of β contains a correlated record $r'_+(r'_-)$ for the value of A measured by β. This is the sense in which the record of the measurement of A is intersubjectively agreed.

As so presented, the existential interpretation faces two related objections. The first, "technical" objection concerns the assumption that the quantity $z(t)$ of equation (8) is and remains zero after some time T. In fact, in the model considered, $z(t)$ merely rapidly becomes very small, and remains small for a considerable time, but eventually becomes large again. But even if $z(t)$ were almost instantaneously to become very small, and remain very small for a time of the order of the age of the universe, it would not be strictly zero during this period. Hence the density operator representing the state of $\sigma \oplus \alpha$ is not quite diagonal in the $\chi_{\pm}$ basis, and the correlation between relative states of α and of $\sigma \oplus \epsilon$ differs slightly but significantly from that described in equation (9). Thus relative to "observer-memory-state" χ^{α}_{+} of α (taken to record the value $a_{\pm}$ for the measured value of A) the quantity A will not have a sharp value during this period. Similarly, relative to "observer-memory-state" χ^{α}_{+} of α, the "observer-memory-state" χ^{β}_{+} of β will contain no record at all for the value of A measured by β. It seems that measurement records are in fact neither faithful nor intersubjectively agreed in the model.

But perhaps it is wrong to think that only the relative state χ^{α}_{+} of α contains a record that the measured value of A was a_+: perhaps any relative state of α that is sufficiently close to χ^{α}_{+} also contains that record? If one makes this weaker assumption about "observer-memory-states," then, if $z(t)$ were almost instantaneously to become very small and remain very small for a long period of time, then throughout that period of time some "observer-memory-state" of α would record the measured value a_+ of A while the correlated relative state of the rest of the universe would make it true that the value of A was indeed a_+. Similarly, loosening the requirements on "observer-memory-states" of β might make it possible to maintain intersubjective agreement between records associated with correlated "observer-memory-states" of α and β, if $z(t)$ were merely to become very small almost instantaneously, and then remain very small for a long period of time. The fact that faithfulness and intersubjective agreement eventually break down would not pose any great difficulty: no records are permanent!

Whether the response of the last paragraph is satisfactory depends, of course, on whether or not it is true that any relative state of α that is sufficiently close to χ_+^α contains a record that the measured value of A was a_+. This raises the second, more fundamental, objection to the existential interpretation: no clear sense has been given to the claim that *any* relative state of an apparatus like α contains a record of the value of the observable measured on a system like σ. Calling such relative states "observer-memory-states" does not answer this objection. In what sense are such states states of an observer's memory? According to Zurek,

> Memory is the stable existence of records—correlations with the state of the relevant branch of the universe. The requirement of stable existence and the recognition of ultimate interdependence between the identities of the observers (determined in part by the physical states of their memories) and their perceptions define the *existential interpretation* of quantum mechanics. (1993b, 88)

Elsewhere Zurek says,

> The perception of the unique outcome is caused by the "redefinition" of the state of the observer, by the change of the observer's identity, which is described in part by the update of the records acquired through the interaction with the measured system. (1993a, 311)

I believe I am not alone in finding such passages obscure. I offer the following interpretation in an exploratory spirit, even though I am far from confident of its correctness, in the hope that, by correcting any misunderstanding it reveals, proponents of the decoherence approach will be able to make their views clearer to skeptics such as myself.

It is natural to assume that in a model such as that presently under discussion, the physical system α plays the role of observer (either a conscious physicist or an unconscious information-processing recording device). This assumption, however, makes it hard to see how the identity of the observer could in any way depend on the relative state of α, or on what that state is taken to record. But only if the observer's identity literally depends on the relative state of α is it possible to resolve the otherwise inescapable contradiction between the observer's record of the measurement being simultaneously both r_+ and r_-. Perhaps, then, there are two distinct observers after time T, at least on the assumption that the quantity $z(t)$ of equation (8) is zero after that time: one corresponding to each of the relative states χ_+^α and χ_-^α. If so, then these two observers cannot both be identified with the single physical system α (and I see no suggestion that the existential interpretation postulates any splitting of physical systems, unlike some "many-worlds" versions of Everett-style interpretations such as that of DeWitt [1970]). Now there is no plausibility to the suggestion that the observers are simply identical to the relative states χ_+^α and χ_-^α. Whatever an observer is, it is certainly not a vector in a Hilbert space, and in any case, this suggestion fails to explain how the observers' identities could be redefined, or change, by any updating of "memory" records, since χ_+^α and χ_-^α are relative states of α at all times.

One possibility is to view the existential interpretation as a variety of "many-minds" interpretation. On this view (see Albert and Loewer [1988]), observers are identified with conscious minds ("observer-minds"), more than one of which is associated with each physical system that is capable of conscious thought. Provided that α is such a system, then after T, assuming that the quantity $z(t)$ of equation (8) is zero after that time, the mental state of some of these observer-minds is represented by relative state χ_+^α: these contain a memory that the outcome of the measurement was a_+. The mental state of other observer-minds is represented by relative state χ_-^α: these contain a memory that the outcome of the measurement was a_-. This would give clear sense both to the idea that relative states partly define the identity of observers (one reason this definition is only partial is that qualitatively identical relative states of distinct physical observer-systems correspond to different observer-minds) and to the idea that an observer's identity is redefined by the incorporation of additional memory records (two observer-minds associated with a particular physical observer-system may contain qualitatively identical mental records up to a certain time, even though their mental records subsequently diverge). But decoherence theorists offer no explicit endorsement of such a "many-minds" view, and instead tend to treat all information-processing recording devices on a par, irrespective of whether or not they are capable of consciousness.

There is a way to treat conscious and unconscious observers on a par while still making sense of these two ideas. Suppose that each observer (conscious or unconscious) is to be identified neither with a physical observer-system nor with a purely mental observer-mind, but rather with a categorically distinct entity I shall call an *observer-process* (Healey [1984]). An observer-process is an object whose successive (absolute) states are defined by a particular temporal sequence of relative states of a particular observer-system. Even though an observer-process is not itself a (quantum-mechanical) physical system, it is an enduring physical object, since it has physical properties at any time; these include at least those properties that follow from application of the eigenvalue-eigenstate link to a relative state of its associated observer-system at that time. More than one observer-process is associated with each physical observer-system; in fact, to allow for the possibility of measurements with an infinite number of possible outcomes, there must be an infinite number of observer-processes associated with at least some observer-systems. Some, but not all, observer-processes are conscious. Presumably, no observer-process associated with the Geiger counter in the Schrödinger cat thought-experiment is conscious, and an observer-process associated with the physical cat will be conscious at its conclusion only if it is associated with a live rather than a dead cat! The physical states of observer-processes, whether conscious or unconscious, may incorporate records of measurement-type interactions that their associated observer-systems have previously undergone with other quantum systems. In the example, an observer-process corresponding to χ^{α}_{+} has the property $R = r_{+}$, which serves as a record that the measured value of A was a_{+}, while an observer-process corresponding to χ^{α}_{-} has the property $R = r_{-}$, which serves as a record that the measured value of A was a_{-}. These records partly define the identities of the observer-processes that contain them insofar as these different records serve to distinguish, and partially to individuate, the two observer-processes (although their prior histories may have been qualitatively identical, thus permitting no such discrimination). Moreover, this "definition" becomes progressively more complete as the records incorporated in the various observer-processes associated with a given observer-system come to diverge, as decoherence following successive measurement-like interactions makes the observer-processes "branch" away from each other as their states change.

Does accepting this explication enable one to overcome objections to the existential interpretation? The introduction of a new category of observer-processes raises a new objection—that of ontological profligacy. One might justly complain that this is not a hidden-variable but a hidden-entity interpretation, but this objection may not be fatal. Ockham only cautioned us against introducing entities beyond necessity: perhaps the necessity of solving the quantum measurement problem is sufficiently pressing to justify the postulation of observer-processes.

Other objections are not so readily deflected. First there is the "technical" objection that even in the oversimplified model considered earlier decoherence is never complete. This seems to imply that for an observer-process associated with state χ^{α}_{+} of α (taken to record the value a_{+} for the measured value of A) the quantity A will not in fact have a sharp value during this period; nor will any observer-process associated with state χ^{β}_{+} of β contain any record at all for the value of A measured by β. One response to this objection would be to allow an observer-process to be associated not just with a single state like χ^{α}_{+} but with a "cone" of states centered on χ^{α}_{+} in the Hilbert space of α, so that any state in that cone would, for that observer-process, correspond to the property $R = r_{+}$, which serves for that observer-process as a record that the measured value of A was a_{+}. But any choice of exact dimensions for such a cone would seem quite arbitrary, corresponding to no clear criterion for the individuation of observer-processes, while any unclarity in their criteria of individuation can only increase the already considerable skeptical doubts about their postulation.

A more fundamental objection to the existential interpretation is that even if that interpretation succeeds in accounting for the observed determinateness of the outcomes of quantum measurements, it does so only at the cost of making it impossible to understand the significance of the probabilities quantum mechanics predicts for each particular determinate outcome. For a probabilistic prediction to be significant, it must concern a situation in which there is some set of possible outcomes, one and only one of which will in fact occur. But according to the existential interpretation there is no sense in which one outcome occurs to the exclusion of all others: every possible outcome of a quantum measurement will occur in some "branch" and be recorded in the memory of an observer-process associated with that "branch." One may seek to avoid this contradiction by understanding the quantum probabilities to be conditional on the history recorded in the memory-states of some particular observer-process. They would then specify the chance that an observer-process with that history would come to incorporate one rather than some other record of the current quantum measurement in its memory-states following the current quantum measurement. But even if one completely specifies the history recorded in the memory-states of an observer-process, every extension of this history by a recording of some possible result of the current quantum measurement will come to be realized in the memory-state of some observer-processes with exactly that history. Hence there is no room for a statement of chance.

If they are to be intelligible, the probabilities must be understood epistemically, as arising from ignorance: given a complete qualitative specification of the history recorded in the memory-states of some particular observer-process, the quantum probabilities specify the likelihood that it is one of the observer-processes whose future memory-states will incorporate this rather than that outcome of the current measurement. Now epistemic probabilities are quite familiar in science. The probabilities associated with different outcomes of a "game of chance" such as roulette arise from ignorance of the exact present state of a system, on the assumption that the laws governing the development of that state are strictly deterministic. A popular view of probabilities in classical statistical mechanics is that they arise because of our inevitable ignorance of the exact microstate of a macroscopic system such as a gas. But in such cases there is at least some historical (indeed present) difference between the exact but unknown states, so that the introduction of probability is intelligible as an attempt to represent the extent of our uncertainty concerning what the state of the system is. But according to the existential interpretation, there is no such difference in the case of quantum mechanics: there may be many distinct observer-processes whose histories prior to the current quantum measurement are qualitatively identical. It follows that there is nothing any observer-process could do to remedy its ignorance of its own identity, and this identity is also inaccessible to all other observer-processes. Such irremediable ignorance has no analog in other circumstances in which epistemic probabilities are featured. For each observer-process, the questions "Who am I?" and "What will happen to me?" collapse into one.

Here is a final, radical reinterpretation of the significance of "probability" within the existential interpretation.[8] Suppose that when an interaction with an observer-system results in decoherent "branches," each associated with different observer-processes, an original observer-process is identical to none of the resulting observer-processes, even though many of these are qualitatively very similar to the original observer-process. A conscious observer-process faced with the prospect of such an interaction might reason as follows. While I myself shall not exist after the coming interaction, I shall leave many "survivors," all qualitatively psychologically similar to me, all of whose mental states will be psychologically continuous with mine. This psychological closeness provides me with grounds for caring as to what will happen to my survivors, for just the same reasons that all of us care about what will happen to our own future self in what we take to be the normal situation.[9]

Now suppose one apportions one's degree of concern for survivors associated with each branch in accordance with the corresponding Born probability coefficient. These "caring coefficients" may be used as weights in a calculation of the total utility of all one's survivors following the coming interaction. This total utility is just what one should consider when determining the appropriate attitude toward this interaction (for example, in making a choice between two alternative interactions). Suppose you are offered a choice between two interactions, each with two possible outcomes, A and B: one observer-process will experience outcome A, the other will experience outcome B. In either case, an observer-process experiencing outcome A will receive one million dollars, while an observer-process experiencing outcome B will receive nothing. In the first interaction, there is a 90 percent Born probability associated with outcome A and 10 percent with outcome B; in the second, there is a 10 percent Born probability associated with outcome A and 90 percent with outcome B. You will receive nothing, since you will no longer exist whatever happens. But you should still prefer the first interaction to the second, since in the first interaction the survivor you care more about would receive the larger sum, while in the second your least favored survivor would receive the larger sum.

This reinterpretation of the Born coefficients may be consistent, but it fails to square with the intuition that my degree of caring for my survivors should depend only on the degree to which each is similar to and continuous with me (especially with regard to psychological features such as character and "memories"). But if one respects this firm intuition, then it appears that there will be many cases in which it dictates that I should care equally for different survivors, even though these survivors are "located" on decoherent branches associated with quite different Born probabilities. The reinterpretation fails to make good sense of the Born probabilities.

It may be that none of these objections to the existential interpretation will prove insuperable. Perhaps that interpretation can be modified, or better explicated, or some related interpretation can be offered, so that such objections can be answered or simply rendered irrelevant. But without further elaboration I fail to see how applying decoherence in the context of an Everett-style relative-state interpretation of quantum mechanics leads to any satisfactory solution to the quantum measurement problem.

5. "Modal" Interpretations

In recent years, a number of philosophers have come to believe that the quantum measurement problem may be solved not through technical developments or novel applications of the quantum formalism, but rather by the adoption of some alternative interpretation of the significance of the quantum state. Although there is no unanimity on exactly how this should be done, I believe that there is widespread agreement that one standard interpretative assumption should be modified or abandoned, namely, the eigenvalue-eigenstate link. Recall that according to this assumption dynamical variable (observable) A has a sharp (real) value a on a system just in case the quantum state of that system may be represented by a state vector that is an eigenvector of the operator $\hat{A}$ corresponding to A. One key tenet of Copenhagen orthodoxy is that quantum mechanics is complete, and adopting the eigenvalue-eigenstate link is one way to maintain the completeness of quantum mechanics. But, as the earlier exposition made clear, the eigenvalue-eigenstate link is a key assumption in the usual chain of reasoning that leads to the quantum measurement problem. On a view like that of Einstein (who rejected the eigenvalue-eigenstate link) quantum mechanics may never face a measurement problem in the first place, because observables may have sharp values even in nontrivially superposed states. Such views have come to be known (rather unfortunately, in my view) as *modal interpretations*. Van Fraassen (1973) first introduced the term "modal" to describe an interpretation according to which a quantum system must have every quantum dynamical property to which its quantum state assigns probability 1, but actually has additional quantum-dynamical properties to which its quantum state assigns probability between 0 and 1. While any such interpretation conflicts with the eigenvalue-eigenstate link, there are interpretations (such as my [1989]) that, although they conflict with the eigenvalue-eigenstate link, do not conform to Van Fraassen's intended usage of the term modal because they deny that a system must have every quantum-dynamical property assigned probability 1 by its quantum state (while acknowledging, of course, that the system is certain—with probability 1—to reveal that property if it is observed).

Any view that attributes sharp values to observables in nontrivially superposed states is automatically suspect. Strong arguments exist, of course, against hidden-variable theories, stemming from the work of Gleason, Kochen and Specker, and Bell. It is therefore important to appreciate how it is possible to reject the eigenvalue-eigenstate link without proposing a theory of a kind that conflicts with any of the famous no-hidden-variable results.

One kind of hidden-variable theory attempts to understand quantum probabilities as measures over a set of dispersion-free states—states in which all observables have precise values, indeed the very values that a reliable measurement would reveal, if it were carried out. Largely algebraic arguments, especially those derived from the work of Gleason and of Kochen and Specker, are widely, and in my view correctly, regarded as refuting any such hidden-variable theory.

A different, and historically more important, motive for proposing a hidden-variable theory has been the desire to restore determinism, and perhaps the most famous hidden-variable theory (that of Bohm) does just that. The idea here is that a complete specification of the state of an individual quantum system would yield a definite (that is, nonprobabilistic) prediction of the value to be found in any measurement of any quantum observable on that system. As Bell and others have pointed out, such a theory may be contextual, thereby evading the negative conclusions of the largely algebraic arguments. But as Bell famously showed, any such deterministic hidden-variable theory must be nonlocal, in the sense that the result of a measurement on one subsystem must depend on what, if anything, is measured (possibly at spacelike separation) on another subsystem of a compound system, whose components are both physically and spatially separated.

Proponents of a modal interpretation need share neither of the motives behind hidden-variable theories of the types that are ruled out by arguments such as these. Such an interpretation requires only that enough observables have sharp values to solve the measurement problem. Although it is necessary to secure sharp "pointer readings," it is certainly not necessary to suppose that there are dispersion-free states, that an electron has a definite trajectory in the two-slit experiment, or that quantum systems have values for position and momentum that conflict with the indeterminacy relations. Nor must a modal interpretation assume that whatever sharp values there are permit the restoration of determinism, or even a statement about the probabilities of results of future quantum measurements on a system more specific than that given by the Born rules. Of course, the details of a view, not the motives of its proponents, determine its consistency with no-hidden-variable results, and recent work has turned up serious problems for some modal interpretations (Bacciagaluppi [1996]). I cannot go into these problems here, and will simply assume that a modal interpretation of the kind I shall discuss can overcome them.

Three varieties of modal interpretation may be distinguished in the recent foundational literature. Each variety offers an account of a quantum measurement that treats it as a dynamical interaction between quantum systems that proceeds in conformity with the Schrödinger equation.

The interpretation of Bub (1996, 1997) relies on an initial specification of some privileged observable R on some class of quantum systems, conceived as measuring apparatus: R is privileged by virtue of the fact that its value is always sharp. While Bub explicitly draws a parallel between his interpretation and Bohm's theory, in which the observable *position* always has a sharp value, he does not go so far as to require that R be a position observable or constructed out of such variables (for instance, as representing a "position" in configuration space specifying the locations of all particles in some set). The eigenvalue-eigenstate link therefore fails for observable R. Bub offers an account of a quantum measurement as an interaction between an object system and an apparatus, which effects a correlation between orthogonal vectors in the object Hilbert space and corresponding eigenvectors of $\hat{R}$ in the apparatus space. Such an interaction is interpretable as a measurement of the value of an observable, on the object system, of whose associated operator in the object Hilbert space these orthogonal vectors are eigenvectors. Consistent with this interpretation, Bub allows the option of a further violation of the eigenvector-eigenvalue link, to permit the assignment of a sharp value also to the measured observable, namely, the eigenvalue corresponding to that observable's eigenvector that is correlated with the (sharp) eigenvalue of $\hat{R}$ that results from the interaction. Thus, suppose that an interaction between an "object" system σ and an "apparatus" system α (initially in the "ready-to-measure" state

χ_0^g) satisfies equation (1). If the initial state of σ is a nontrivial superposition, then the interaction proceeds as in equation (2). But the right-hand side of that equation is now interpreted as follows: even though their joint state is an "entangled" superposition, the privileged observable R has some sharp value r_i on α, and the interaction may be taken to constitute a measurement of the quantity A on σ with result a_i. Bub (1996) appears noncommittal as to whether or not A *has* value a_i at any time before, during, or after this interaction, but he takes the attribution of r_i to R to be the key to solving the measurement problem; this ensures a determinate outcome for a quantum measurement.

Van Fraassen, who coined the phrase 'modal interpretation,' has been elaborating and advocating his own version of such an interpretation for twenty years. The most complete presentation appears in van Fraassen (1991). Van Fraassen's modal interpretation treats all observables "democratically"; none is singled out as privileged in that it always has a sharp value. Though in this respect more liberal, his interpretation is in another way more conservative: only at the conclusion of a very special type of quantum-mechanical interaction is any particular observable required to have a sharp value. Van Fraassen spells out conditions on such interactions, which he then takes to define the class of measurement interactions. The conditions on the Hamiltonian for such an interaction suffice, within the interpretation, to pick out certain observables on one of the interacting systems—the "apparatus" system—that have sharp values at its conclusion. These conditions also define a mutually commuting set of observables on the other system, such that the interaction counts as a simultaneous measurement of all of them. The interpretation does attribute sharp values to other observables, both in measurement contexts and outside them. The rule governing these further value attributions to a system is generally quite permissive; it picks out no particular observable as having a sharp value. Basically, the only constraint on these values is that there be *some* vector from the image space of the reduced density operator representing the statistical state of the system such that these values are consistent with the application of the eigenvalue-eigenstate rule to that vector. But even this permissive rule attributes values to observables on a subsystem inconsistent with application of the eigenvalue-eigenstate rule to the vector representing the "entangled" quantum state of the total system of which it is a component. Van Fraassen addresses the semantic aspect of the measurement problem by offering a precise, quantum-mechanical characterization of a type of interaction that is to count as a measurement, and he addresses its consistency aspect by showing that, on his interpretation, such an interaction always results in a final state of a quantum-mechanical "apparatus" system in which its quantum-mechanical "pointer position" observable has a sharp value. An interaction between σ and α counts as a measurement of A on σ by α if (i) the total Hamiltonian is such that equation (1) is satisfied for arbitrary χ_i^σ; and (ii) the initial quantum state of α is χ_0^α.

The final joint state on the right-hand side of equation (2) is to be interpreted as follows. What van Fraassen calls the dynamic state of α at the conclusion of the interaction is represented by the reduced density operator $W^\alpha = \Sigma\,|c_i|^2 P_i^\alpha$, where P_i^α projects onto the subspace spanned by χ_i^α. It is then taken to follow from the fact that this is its dynamic state at the conclusion of a measurement-type interaction that what van Fraassen calls the value state of α is then χ_i^α, for some i, with probability $|c_i|^2$. Applying the eigenvalue-eigenstate link to this value state (rather than to the state of $\sigma \oplus \alpha$, or to the dynamic state of α), van Fraassen concludes that, with probability $|c_i|^2$, the value of R at the conclusion of the measurement is r_i. Similarly, since the final dynamic state of σ is $W^\sigma = \Sigma\,|c_i|^2 P_i^\sigma$, its final value state is χ_j^σ (with probability $|c_j|^2$), and so, with probability $|c_j|^2$, the value of A at the conclusion of the measurement is a_j. (Interestingly, van Fraassen seems to be of two minds about whether or not further to require that $i = j$.)

A third variety of modal interpretation was first proposed by Kochen (1985). Other variants have been developed by Dieks (1989a,b; 1994a,b) and Healey (1989; 1993a,b; 1995). Like van Fraassen, and unlike Bub, this variety treats all quantum-dynamical variables democratically; which observables are taken to have sharp values in a given situation depends on the appropriate quantum-mechanical description of that situation, rather than on some universal external decree. But like Bub, and unlike van Fraassen, sharp values are ascribed to particular observables in circumstances other than those obtaining at the end of a measurement interaction. Modal interpretations of this third variety apply a simple consequence of the polar decomposition theorem to determine which observables have sharp values.

Biorthogonal Decomposition

Let Ψ be a vector in the tensor product $\mathcal{H}^\alpha \otimes \mathcal{H}^\beta$ of two Hilbert spaces $\mathcal{H}^\alpha$, $\mathcal{H}^\beta$. Then there exist sets of orthonormal vectors $\{\chi_i^\alpha\}$, $\{\chi_i^\beta\}$ in $\mathcal{H}^\alpha$, $\mathcal{H}^\beta$ respectively such that Ψ may be written as

$$\Psi = \textstyle\sum_i c_i(\chi_i^\alpha \otimes \chi_i^\beta) \tag{13}$$

Moreover, the sets $\{\chi_i^\alpha\}$, $\{\chi_i^\beta\}$ are unique (up to phase) if and only if $|c_i|^2 \neq |c_j|^2$ for $i \neq j$.

The interpretations of Kochen, Dieks, and Healey do not apply this result in quite the same way, corresponding to significant further differences between these interpretations. To proceed further it is necessary to consider each of these interpretations individually, but to treat all of them adequately here is impossible. I therefore restrict my attention to the treatment of measurement within the interpretation with which I am most familiar, namely my own.

Here is (a slight simplification of) a model presented in Healey (1989). Consider a quantum-mechanical interaction between two quantum systems σ and α governed by a Hamiltonian that would induce the following time evolution over the course of the interaction, for each normalized eigenvector χ_i^σ of an operator $\hat{A}$:

$$\chi_i^\sigma \otimes \chi_0^\alpha \Rightarrow \xi_i^\sigma \otimes \chi_i^\alpha, \tag{14}$$

where $(\chi_i^\sigma, \chi_j^\sigma) = (\xi_i^\sigma, \xi_j^\sigma) = (\chi_i^\alpha, \chi_j^\alpha) = \delta_{ij}$. I call such an interaction *M-suitable for A*. Not every interaction governed by this Hamiltonian would count as a measurement of A, however: the system α must be in the correct state as the interaction begins, and α must also be the right kind of system. To explain these further restrictions, I need to say more about the structure of the interpretation of quantum mechanics presented in Healey (1989).

On this interpretation, the quantum-dynamical properties of every quantum system, whether simple or compound, are specified by what I call its *dynamical state*. If a system has no other subsystems, then its dynamical state is in turn specified by its *system representative*, a vector (or more generally subspace) in the system's Hilbert space; the system has a dynamical property just in case the subspace corresponding to that property includes the system representative. The dynamical state of a compound system is partly specified by its system representative, and partly by the system representatives of its component subsystems. The general effect is to make the properties of a compound system somewhat more definite than its system representative would lead one to expect.[10]

Certain quantum systems may, in certain circumstances, also be ascribed quantum states. The quantum state of a system does not describe its properties; instead, it specifies the chances that future measurement-type interactions between that system and another system will produce a dynamical state in the latter that includes one dynamical property from a set of properties, each member of which may be taken to indicate a different result of the measurement. Sometimes, but not always, a system may be ascribed its system representative as its quantum state.[11] In this case, the same vector does double duty: it (helps to) specify what properties the system has, and also provides a quantitative measure of the dispositions for that system to produce each of a set of different possible results on measurement. But even in this case, the dynamical state and quantum state are conceptually distinct: the former records intrinsic, categorical properties of the system, while the latter specifies relational, probabilistic dispositions involving not only the system but also other systems correlated with it, or with which it is to interact.

If an interaction between σ and α is to count as a simple M-type interaction, then the initial system representative of α must be χ_0^α. A further technical restriction is also required, this time on the structure of the dynamical properties of α, which is spelled out, purely in terms of the Hilbert spaces of α and its components, in section 3.1 of my 1989 book.

Now suppose that α meets this last restriction, and that its initial system representative is indeed χ_0^α. Suppose that σ is a simple quantum system that initially has value a_i for observable A. Then the initial system representative of σ is χ_i^σ. If we further assume that σ and α are initially uncorrelated, then the system representative of $\sigma \oplus \alpha$ will be the tensor product vector $\chi_i^\sigma \otimes \chi_0^\alpha$. On my interpretation, the Schrödinger equation is grounded on the fact that the system representative of an isolated system evolves in accordance with the appropriate unitary time-evolution operator. Assuming that the system $\sigma \oplus \alpha$ is isolated from all external disturbances, equation (14) will then correctly describe the evolution of its system representative over the course of an interaction that is M-suitable for A on σ.

The interpretation postulates a relation between the system representative of a compound system and the system representatives of its component subsystems. This relation implies that the system representatives of σ and α after the interaction represented by (14) will be ξ_i^σ and χ_i^α respectively. It follows that immediately after this interaction the pointer reading quantity R on α has acquired the sharp value r_i. But it also follows that observable A on σ does *not* have a sharp value, unless $\xi_i^\sigma = \chi_j^\sigma$ (for some j). This is not a problem for the present interpretation, on which quantum mechanics does not specify the chance that an observable will acquire a definite value on measurement, but rather the chance that an appropriate "apparatus" device will record a definite value, irrespective of whether or not the system then has that value. In the present case, α is certain to record value a_i, and this is indeed the value of observable A on σ immediately prior to the measurement, although A has this value on σ after the interaction only if the measurement is nondisturbing, so that $\xi_i^\sigma = \chi_i^\sigma$.

Now suppose instead that the system representative of σ immediately before the interaction is given not by an eigenvector of $\hat{A}$, but by a superposition of such eigenvectors. It then follows from the linearity of the time-evolution law for system representatives underlying the Schrödinger equation that the initial system representative of $\sigma \oplus \alpha$ evolves as follows under the influence of the same Hamiltonian:

$$\left[\sum_i c_i \chi_i^\sigma\right] \otimes \chi_0^\alpha \Rightarrow \sum_i c_i [\xi_i^\sigma \otimes \chi_i^\alpha], \; \sum_i |c_i|^2 = 1. \tag{15}$$

In this case A has no sharp value on σ before the interaction. What quantities have values after the interaction? To answer this question, it is necessary to specify the condition that relates the system representative of a compound system like $\sigma \otimes \alpha$ to those of its subsystems, σ and α. According to the Subspace Decomposition Condition, if the system representative of $\sigma \oplus \alpha$ is a vector whose unique biorthonormal decomposition onto the Hilbert spaces of components σ and α appears on the right-hand side of (15), then, with probability $|c_i|^2$, the system representatives of σ and α are some correlated pair $\langle \xi_i^\sigma, \chi_i^\alpha \rangle$. As a special case, when the system representative of $\sigma \oplus \alpha$ is given by the right-hand side of equation (14), the system representatives of σ and α are $\langle \xi_i^\sigma, \chi_i^\alpha \rangle$, with probability 1. It then follows that after the interaction described by (15), the pointer reading quantity R on α has now acquired some sharp value r_i. It also follows that observable A on σ has acquired the correlated value a_i if (but only if) $\xi_i^\sigma = \chi_i^\sigma$.

As a general model of measurement interactions, equations (14) and (15) are clearly vastly oversimplified. Albert and Loewer (1990, 1993) and Elby (1993) objected that if one attempts to make this model more realistic by removing certain extreme idealizations, then the measurement problem reemerges once more. Specifically, they argued that real measurements are prone to error, and that a similar application of the interpretation of Healey (1989) to models of error-prone measurements fails to explain why these, too, always have definite (though occasionally misleading) outcomes. The objection is that real quantum-measurement interactions are *not* appropriately modeled by (15), and so the definiteness of outcomes has *not* in fact been derived. They further argued that if one replaces (15) by a more realistic model of a quantum measurement interaction, the interpretation implies the *in*definiteness of measurement outcomes in most, if not all, circumstances; and that consequently the interpretation of Healey (1989) fails to solve the measurement problem. They based this stronger conclusion on a particular model of a quantum measurement interaction, designed to take account of the fact that many, if not all, real quantum measurements are error-prone. That is to say, in some small fraction of cases the apparatus records the wrong value.

That model is basically as follows:

$$\left[\sum_i c_i \chi_i^\sigma\right] \otimes \chi_0^\alpha \Rightarrow \sum_{ij} c_{ij} [\xi_i^\sigma \otimes \chi_j^\alpha], \tag{16}$$

where $\Sigma_{ij} |c_{ij}|^2 = 1$; $|c_{ij}|^2 \ll 1$ if $i \neq j$; $c_{ii} \approx c_i$. (Both Albert and Loewer and Elby restricted their attention to the "nondisturbing" case in which $\xi_i^\sigma = \chi_i^\sigma$.) It is always possible to find a set of vectors θ_i^α such that equation (16) may be re-expressed as follows:

$$\left[\sum_i c_i \chi_i^\sigma\right] \otimes \chi_0^\alpha \Rightarrow \sum_i c_i [\xi_i^\sigma \otimes \theta_i^\alpha], \tag{17}$$

where (in general) the vectors θ_i^α almost, but not quite, form an orthonormal set.

Neither the right-hand side of (16) nor the right-hand side of (17) generally constitutes a biorthonormal decomposition of the system representative of $\sigma \oplus \alpha$. The subspace decomposition condition therefore implies that neither χ_i^α nor θ_i^α is a possible system representative of α. The actual system representative of α will then correspond to a sharp value for some quantity other than R, while the value of R itself is rendered unsharp (that is, for every real number r, it is false that the value of R is r). It is therefore a consequence of my interpretation that, just after an interaction modeled by (16), the quantity R will fail to have a sharp value. Moreover, for some sets of coefficients $\{c_i\}$ the actual system representative of α at the conclusion of the interaction is not even close (in the Hilbert space norm) to any eigenvector of $\hat{R}$. In that case, the quantity that does then have a sharp value will, in an intuitive sense, be very different from R.

Albert and Loewer argued as follows. While my interactive interpretation is able to secure a definite outcome for an ideal measurement (satisfying [15]), or perhaps just its restriction to the "nondisturbing" case in which $\xi_i^\sigma = \chi_i^\sigma$), real measurements are not ideal. Instead, real measurements are error-prone: even when the measured observable has some sharp value, the pointer may (in a small fraction of the cases) come to record some quite different value. They claimed that (16), rather than (15), is an appropriate model for many, if not all, real quantum measurement interactions; they concluded that since an interaction modeled by (16) almost never produces a definite outcome on my interpretation, that interpretation has failed to solve the measurement problem. Elby (1993) agreed with the thrust of Albert and Loewer's argument, and tried to strengthen it by arguing that no quantum measurement could satisfy (15)—that error-free measurements are physically impossible.

I attempted to answer this objection in Healey (1993a,b) by developing simple alternative models of error-prone measurement interactions that, on my (1989) interpretation, do yield sharp values of pointer-reading observables even for nontrivial initial superposed states. But these models were able to purchase this desirable consequence only by making very special, and rather unnatural, assumptions about the form of the interactions involved in error-prone measurements. More recent work by Bacciagaluppi and Hemmo (1994a) has shown how such artificial assumptions may be replaced by very natural assumptions about the effects of environmental decoherence on the state of an "apparatus" system involved in a quantum measurement. That approach is further explored in Healey (1995). The next section reviews these and other attempts to apply decoherence results in the context of a "modal" interpretation to solve the quantum measurement problem.

6. Decoherence within a "Modal" Interpretation

None of the treatments of measurement presented in the previous section included explicit models for interactions with the environment during or after a measurement interaction. But clearly such interaction cannot be excluded in any actual measurement, and so must itself be modeled in a more realistic quantum-mechanical treatment of measurement. What happens when a "modal" interpretation attempts such modeling?

To my knowledge, van Fraassen's modal interpretation has not been significantly developed in this direction (but see his comments on an early paper by Zurek in van Fraassen [1991, 216–18]). He appears not to regard such development as required to solve what he takes to be the heart of the measurement problem—that is, "the general problem of drawing a consistent general recipe for probability assignments from quantum theory" (285). In terms of my discussion in section 2, van Fraassen here acknowledges only the first and second aspects of the quantum measurement problem—the semantic and consistency aspects. A fully adequate solution to the quantum measurement problem must also address its third and fourth aspects. In order to explain why we believe quantum mechanics to be an empirically adequate theory it must present realistic models of those measurements we take to confirm the Born rules, and in order to reconcile quantum mechanics with our everyday experience of a classical world it must account for (our experiences of) the definite positions of pointers, whether or not these are used to perform some quantum measurement.

Bub (1996) also sees his modal interpretation of quantum mechanics as solving the measurement problem largely without reference to environmental interactions. At one point he does make a passing appeal to decoherence to explain why the privileged observable R is unlikely to fluctuate much stochastically between values associated with different values of a measured observable in a repeated measurement as a consequence of interference effects. But he stresses that this appeal is quite unobjectionable because the events in question are always determinate and do not depend for their determinateness on the smallness of certain probabilities. Rather, on his view, decoherence explains why certain events that we would regard as anomalous will occur very rarely.

I see a much more substantial role for decoherence within "modal" interpretations that appeal to biorthogonal decomposition. Within such interpretations, decoherence can explain how the sharpness of values of quantum observables implied by the interpretation can account for our experiences of definite "pointer positions" both following actual quantum measurements (error-prone or not) and in other circumstances. The rest of this section shows this for simple models in the context of the interpretation of Healey (1989).

Suppose an interaction between a system σ and apparatus α is governed by a Hamiltonian that would give rise to an evolution of the form (14) for a quantity A, that the interaction is M-suitable for A, and that $\hat{A}$ has two nondegenerate eigenvalues a_+, a_- on the two-dimensional Hilbert space of σ. Suppose further that $\chi_0^\alpha = \chi_+^\alpha$, and that an observable R on α that I shall call the exact pointer reading quantity has nondegenerate eigenvalues r_+, r_- (with corresponding eigenvectors χ_+^α, χ_-^α) on the two-dimensional Hilbert space of α. (Of course, supposing that the Hilbert space of α is only two-dimensional is a huge oversimplification: the Hilbert space of any realistic apparatus will be infinite-dimensional, and the spectrum of any realistic $\hat{R}$ will likely contain far more than two values.) Then equation (14) may be rewritten as

$$\chi_\pm^\sigma \otimes \chi_+^\alpha \Rightarrow \xi_\pm^\sigma \otimes \chi_\pm^\alpha. \tag{18}$$

Similarly, equation (15) becomes

$$\begin{gathered}[c_+\chi_+^\sigma + c_-\chi_-^\sigma] \otimes \chi_+^\alpha \Rightarrow c_+ \, [\xi_+^\sigma \otimes \chi_+^\alpha] + c_- \, [\xi_-^\sigma \otimes \chi_-^\alpha], \\ |c_+|^2 + |c_-|^2 = 1\end{gathered} \tag{19}$$

It is now necessary to model the effect of the environment on this measurement interaction. In doing so here I shall make a number of simplifying assumptions. First, assume that there is no environmental interaction at all until the measurement interaction (19) is over. Second, assume that the environment interacts only with α, and not with σ. Third, assume that the environment ϵ is modeled as in section 3 by a large number N of two-state systems, the kth of which has a Hilbert space spanned by the basis $\{\kappa_+, \kappa_-\}$. Assume finally that the form of the interaction between α and ϵ is the same as that assumed in section 3. It follows that at a later time t, the system representative of $\sigma \oplus \alpha \oplus \epsilon$ is

$$\begin{aligned}\Psi(t) = c_+\chi_+ \Pi_k \otimes [c_+^k e^{ig_k t}\kappa_+ + c_-^k e^{-ig_k t}\kappa_-] \\ + c_-\chi_- \Pi_k \otimes [c_+^k e^{-ig_k t}\kappa_+ + c_-^k e^{ig_k t}\kappa_-]\end{aligned} \tag{20}$$

where

$$\chi_{\pm} = \xi^{\sigma}_{\pm} \otimes \chi^{\alpha}_{\pm}. \qquad (21)$$

Then the quantum state of $\sigma \oplus \alpha$ is represented by the reduced density operator

$$\rho = |c_+|^2 P_{\chi_+} + |c_-|^2 P_{\chi_-} + z(t)\, c_+ c^*_- Q^{\pm}_{-} + z^*(t)\, c^*_+ c_- Q^{-}_{+} \qquad (22)$$

where

$$z(t) = \Pi^{N}_{k=1} [\cos 2g_k t + i(|c^k_+|^2 - |c^k_-|^2)\sin 2g_k t], \quad Q^{+}_{-}\chi_- = \chi_+, \quad Q^{-}_{+}\chi_+ = \chi_-. \qquad (23)$$

Now if the reduced density operator of $\sigma \oplus \alpha$ were exactly diagonal in the $\chi_{\pm}$ basis at the end of this environmental interaction, then the system representative of α would become either χ^{α}_{+} or χ^{α}_{-}, thus guaranteeing a sharp value for the exact pointer reading quantity R. Moreover, it follows from the stability condition of Healey (1989, 82) that the system representative of α would be χ^{α}_{+} if and only if the system representative prior to environmental interaction were χ^{α}_{+}. To this extent, the interaction between α and ϵ would preserve the value of the exact pointer reading quantity R.

But in fact the reduced density operator of $\sigma \oplus \alpha$ merely extremely rapidly becomes very nearly diagonal in the $\chi_{\pm}$ basis, and the system representative of α merely rapidly approaches either χ^{α}_{+} or χ^{α}_{-}. The effect of the environmental interaction is therefore not to preserve the sharp value of the exact pointer reading observable R, but to destroy it. It seems that this "modal" interpretation cannot solve the measurement problem even for an error-free measurement interaction such as (18).

If the quantity $z(t)$ is very small compared to $\| c^2_+ | - | c^2_- \|$, then the biorthogonal decomposition of the system representative of $\sigma \oplus \alpha \oplus \epsilon$ will be of the form

$$\Psi(t) = \overline{c_+}(t)\, [\overline{\chi^{\alpha}_{+}}(t) \otimes \overline{\chi^{\sigma\oplus\epsilon}_{+}}(t)] + \overline{c_-}(t)\, [\overline{\chi^{\alpha}_{-}}(t) \otimes \overline{\chi^{\sigma\oplus\epsilon}_{-}}(t)] \qquad (24)$$

where $\overline{\chi^{\alpha}_{\pm}}(t) \approx \chi^{\alpha}_{\pm}$, $\overline{c_{\pm}}(t) \approx c_{\pm}$. If N is sufficiently large, decoherence proceeds so rapidly that, unless $|c_+| = |c_-|$, $\overline{\chi^{\alpha}_{\pm}}(t) \approx \chi^{\alpha}_{\pm}$ for very small values of t, and this approximate equality persists for a very long time thereafter. This is true even if $|c_+|$ and $|c_-|$ are extremely close (Bacciagaluppi and Hemmo [1996]). Nevertheless, the exact pointer-reading quantity R will not have a sharp value in $\overline{\chi^{\alpha}_{\pm}}(t)$ unless $\overline{\chi^{\alpha}_{\pm}}(t) = \chi_{\pm}$. Hence the effect of environmental interaction with α will be that, at almost any time after the end of the measurement interaction, the exact pointer reading quantity R fails to have a sharp value.

But this does not show that the measurement fails to have a definite, stable outcome. For the condition on α for its state to show a definite outcome is not that R should have a sharp value, but rather that some *approximate* pointer-reading quantity should have a sharp value. Here, an approximate pointer-reading quantity may be taken to be represented by some operator $\hat{Q}$ on the Hilbert space of α with the same eigenvalues as $\hat{R}$, but whose eigenvectors lie in subspaces that are merely extremely close to those defined by the eigenvectors of $\hat{R}$. If these subspaces are sufficiently close, then an observation that the value of Q is $r_+(r_-)$ will be empirically indistinguishable from an observation that the value of R is $r_+(r_-)$.[12]

Now the $\overline{\chi^{\alpha}_{\pm}}(t)$ are eigenvectors of an operator $\hat{Q}(t)$, with the same eigenvectors as $\hat{R}$; after a very short time t, $\overline{\chi^{\alpha}_{\pm}}(t) \approx \chi^{\alpha}_{\pm}$, and so the eigenvectors of $\hat{Q}(t)$ and $\hat{R}$ will indeed lie in subspaces that are extremely close. It follows that interactions with the environment will extremely rapidly result in the state of α being such that an approximate pointer-reading quantity has a sharp value. That is a condition empirically indistinguishable from the exact pointer-reading quantity R's having a sharp value.

But what about the stability of the value of approximate pointer-reading quantities? If $Q(t_1)$, $Q(t_2)$ are distinct approximate pointer-reading quantities that have sharp values following a measurement interaction, the stability condition of Healey (1989) does not imply that these values are the same. Some further constraint must therefore be assumed to govern the stochastic evolution of dynamical states postulated in Healey (1989) if quantum mechanics is to account for the observed stability of pointer readings. A very plausible constraint will suffice: if Q_r, Q_s are quantities whose associated operators have identical eigenvalues and eigenspaces that are pairwise very close, and the value of Q_r is $r_+(r_-)$ at t, then if Q_s has a value a short time after t, it is almost certain that this value will also be $r_+(r_-)$.

With the addition of such a constraint, quantum mechanics (under the interpretation of Healey [1989]) explains (in the context of this highly simplified model) why the state of α following an error-free quantum measurement interaction would be observed to contain a stable record of the outcome of the measurement. Since $|\overline{c_{\pm}}(t)|^2 \approx |c_{\pm}|^2$, even for very small times t, the probability of recording a particular result rapidly becomes indistinguishable from the quantum-mechanical probability predicted by the Born rules.

This simplified model may now be extended to the case of error-prone measurements. Consider an interaction between systems σ and α of the kind described, governed by a Hamiltonian that would result in the following evolution over the course of the interaction:

$$(c_+\chi^{\sigma}_+ + c_-\chi^{\sigma}_-) \otimes \chi^{\alpha}_+ \Rightarrow c_{++}(\xi^{\sigma}_+ \otimes \chi^{\alpha}_+) + c_{+-}(\xi^{\sigma}_+ \otimes \chi^{\alpha}_-) + c_{-+}(\xi^{\sigma}_- \otimes \chi^{\alpha}_+) + c_{--}(\xi^{\sigma}_- \otimes \chi^{\alpha}_-) \quad (25)$$

where $c_{++} \approx c_+$, $c_{--} \approx c_-$; c_{+-}, $c_{-+} \approx 0$. This models an error-prone measurement of A on σ along the lines of equation (16). Just at the conclusion of this interaction, the interpretation of Healey (1989) implies that R fails to have a sharp value on α. Now it is necessary to take account of the effects of interactions between α and the environment ϵ.

Equation (25) may be rewritten as a biorthonormal expansion

$$\overline{c_+}(0)\ [\overline{\xi_+^\sigma}(0) \otimes \overline{\chi_+^\alpha}(0)] + \overline{c_-}(0)\ [\overline{\xi_-^\sigma}(0) \otimes \overline{\chi_-^\alpha}(0)] \qquad (26)$$

in which $\overline{c_\pm}(0) \approx c_\pm$ and $\overline{\chi_\pm^\alpha}(0) \approx \chi_\pm^\alpha$ unless $|c_+| \approx |c_-|$. If α is now subject to an environmental interaction of the same form as before, then provided that the quantity $z(t)$ is very small compared to $\| c_+^2| - |c_-^2\|$, this will once more produce an evolution toward a state whose biorthonormal expansion is of the form

$$\Psi(t) = \overline{c_+}(t)[\overline{\chi_+^\alpha}(t) \otimes \overline{\chi_+^{\sigma\oplus\epsilon}}(t)] + \overline{c_-}(t)[\overline{\chi_-^\alpha}(t) \otimes \overline{\chi_-^{\sigma\oplus\epsilon}}(t)] \quad (27)$$

where $\overline{\chi_\pm^\alpha}(t) \approx \chi_\pm^\alpha$, $\overline{c_\pm}(t) \approx c_\pm$. If N is sufficiently large, decoherence proceeds so rapidly that, unless $|c_+| = |c_-|$, $\overline{\chi_\pm^\alpha}(t) \approx \chi_\pm^\alpha$ for very small values of t, and this approximate equality persists for a very long time thereafter. This is true even if $|c_+|$ and $|c_-|$ are extremely close.

It follows that, even a very short time t after the conclusion of the interaction between σ and α, some approximate pointer-reading quantity $Q(t)$ will have a sharp value that will serve to record the outcome of the measurement of quantity A on σ, and this will continue to be the case for a very long time afterward. Moreover, the chance that the recorded value is $a_\pm$ is $|\overline{c_\pm}(t)|^2 \approx |c_\pm|^2$. Provided that the dynamics are constrained in the way described earlier, it is almost certain that any other approximate pointer-reading quantity $Q(t')$ that has a sharp value a short time after the conclusion of the interaction between σ and α will record the same value $a_\pm$. This explains in terms of the same simplified model why error-prone measurements are always observed to produce stable records, with statistics that, by careful experimental design, may be made to approximate the quantum-mechanical probabilities arbitrarily closely.

7. Explaining Experience

In order to solve the measurement problem it is necessary to show why quantum measurements are observed to yield unique, determinate results. Consistent with a linear dynamics, there are two general strategies for doing this. The radical strategy denies that quantum measurements yield unique, determinate results, but it then attempts to show why it appears to observers like us that they do; this seems the strategy behind Zurek's existential interpretation. An apparently more conservative strategy attempts to show that measurements do yield unique, determinate results; this is then taken to provide a natural explanation as to why such measurements are observed to yield unique, determinate results. This is the strategy behind "modal" interpretations and was pursued in the previous section in the context of the interpretation of Healey (1989). But for this strategy to succeed, it must make clear the connection between the unique, determinate results of quantum measurements and our observations of these unique, determinate results. Such a connection is trivial in the context of classical physics. There, the determinate pointer position simply causes our experiences of its having a determinate position; a more complete account of this process may be safely left to sciences such as optics, physiology, and neuroscience. Things are not so simple in the different context of quantum mechanics, interpreted along the lines of Healey (1989). The final problem posed for that interpretation by quantum measurements is to explain why we experience apparatus (and macroscopic objects more generally) as having the determinate properties the interpretation says they have.

The key difference from the context of classical physics is that in the present context the "pointer position" does not play a direct role in the dynamical process that gives rise to a final physical state of some physical system or systems that corresponds to an observer's experiential state. It could not play such a role if the observer were not even identified with any quantum system or systems, as in certain Everett-style interpretations. But even when the observer's experiences are modeled in terms of dynamical states of quantum observer-systems (as in the "modal" interpretation presently under consideration), it is the overall dynamical state of the total system of which an observer-system is merely one component that fixes (through application of biorthogonal decomposition in the subspace decomposition condition) what these states might be after the observer-system interacts with the apparatus. This raises the serious worry that the biorthogonal decomposition of the total state may so constrain the dynamical state of an observer-system as to rule out the possibility of interpreting this as corresponding to *any* experience of a determinate "pointer position."

In order to address this worry, it is necessary to consider quantum-mechanical models of an interaction between a physical observer-system and a physical apparatus-system. Here the physical observer-system is a stand-in for either a human observer or some other information-recording device, conscious or nonconscious. The idea is to use such models to show that such a system will almost certainly be put into a state in which it contains a record of or "belief" about a definite result, and that this will almost certainly faithfully correspond to the determinate "pointer reading" of the apparatus system.

A quantum-mechanical observer-system is basically just another apparatus-system, and its observation of an apparatus "pointer reading" is basically just a rather coarse, nondisturbing quantum measurement. To show that observation of a determinate "pointer position" will yield a definite "belief," it is necessary to show that a nondisturbing measurement of the "pointer-reading" quantity on one apparatus by another apparatus will almost certainly give rise to a determinate, confirming record in the latter.

In light of the preceding discussion, it should be clear that such a record cannot be constituted by the stable, determinate value of a single quantum-mechanical observable on the observer-system. There will be no such stable, determinate value on the present interpretation. But for the state of the observer-system to contain a stable record it is not necessary that it possess some stable, determinate value of a unique quantum observable. Consider how a record is contained in a typical information-storage device such as a computer disk, a book, or even the brain of some organism. The device is macroscopic, and the record it contains is quite insensitive to its exact microstate (indeed, the extreme insensitivity of digital information storage devices is one of their chief advantages). A natural way to model such insensitivity at the quantum level is to suppose that the device's Hilbert space $\mathcal{H}$ is spanned by a set of mutually orthogonal subspaces $\mathcal{P}_i$, such that if (but *not* only if) its system representative lies in $\mathcal{P}_i$ the system will be in the ith mesoscopic state. Some of these mesoscopic states represent records that the apparatus indicated a definite outcome, while others do not (perhaps the observer-system is strewn about the universe!). Let the former states correspond to values of i such that $i \epsilon I$. Designating the observer-system as β, let χ_{ij}^{β} represent the jth microstate associated with the subspace $\mathcal{P}_i$ $(i \epsilon I)$; then if its system representative is any superposition of the form $\Sigma_j c_j \chi_{i'j}^{\beta}$, the observer-system β contains a record that the apparatus α with which it has just interacted indicated a definite outcome of its measurement on σ. But what if β's system representative is a superposition containing elements from *distinct* subspaces $\mathcal{P}_i$? It may seem that such states may well result from β's interactions with α, and that in such a state β contains *no* determinate record.

Now it is possible to appeal to decoherence once more, this time to show that the system representative of β will extremely rapidly assume the form $\Sigma_{ij} c_{ij} \chi_{ij}^{\beta}$ where $|c_{ij}|^2 \ll 1$ for $i \neq i'$. This system representative corresponds to a state of β that also contains a record that the apparatus α with which it has just interacted indicated a definite outcome of its measurement on σ. The justification for this last claim is as follows. The content of a record is determined not so much by the intrinsic character of the physical state of a system that bears the record as by the precise functional role of that state. (Think, for example, of the extraordinary variety of physical states an object can be in that are perceptually equivalent with respect to color, given the peculiar features of human color vision, or the radically different ways in which movies and television present perceptually equivalent records.) But as regards their precise functional role, the states $\Sigma_j c_j \chi^{\beta}_{i'j}$ and $\Sigma_{ij} c_{ij} \chi^{\beta}_{ij}$ (where $|c_{ij}|^2 \ll 1$ for $i \neq i'$) are equivalent, since any subsequent interaction with β that effectively "reads" the record contained in its state will have basically the same outcome for both these initial states of β.

To see how this works out in more detail, suppose that the dynamical state produced by environmental decoherence following a measurement interaction by α on σ is that given by equation (28):

$$\Psi(t) = \overline{c_+}(t)[\overline{\chi^{\alpha}_+}(t) \otimes \overline{\chi^{\sigma\oplus\epsilon}_+}(t)] + \overline{c_-}(t)[\overline{\chi^{\alpha}_-}(t) \otimes \overline{\chi^{\sigma\oplus\epsilon}_-}(t)] \quad (28)$$

where $\overline{\chi^{\alpha}_{\pm}}(t) \approx \chi^{\alpha}_{\pm}, \overline{c_{\pm}}(t) \approx c_{\pm}$.

The claim is that observation of α by observer-system β will now almost certainly produce in β a record that the apparatus α indicated a definite result of its measurement of A on σ, and moreover this will record that the result was a_+ just in case α did in fact indicate that the result was a_+. To substantiate this claim it is necessary to model the interaction between β and α by virtue of which β may come to record what α indicates. In the present simple model that interaction may be taken to be as follows:

$$\chi^{\alpha}_{\pm} \otimes \chi^{\beta}_{0} \Rightarrow \chi^{\alpha}_{\pm} \otimes \chi^{\beta}_{\pm}. \quad (29)$$

Note that this interaction has been taken to be both nondisturbing and error-free. No actual interaction will conform exactly to these idealizations, but removing the former would introduce no significant complications, and the treatment of error-prone interactions in section 6 indicates why this is also the case for the latter idealization.

Ignoring all interactions except that between β and α, the resulting dynamical state will then be expressible in the form

$$\begin{aligned}\Psi(t') = {} & \psi^{\sigma\oplus\epsilon}_1(t') \otimes \chi^{\alpha}_+ \otimes \chi^{\beta}_+ + \psi^{\sigma\oplus\epsilon}_2(t') \otimes \chi^{\alpha}_- \otimes \chi^{\beta}_- \\ & + \psi^{\sigma\oplus\epsilon}_3(t') \otimes \chi^{\alpha}_{+*} \otimes \chi^{\beta}_{+*} + \psi^{\sigma\oplus\epsilon}_4(t') \otimes \chi^{\alpha}_{-*} \otimes \chi^{\beta}_{-*}\end{aligned} \quad (30)$$

where the respective χ are pairwise orthogonal, the ψ are almost pairwise orthogonal, $|\psi_1|^2 \approx |c_+|^2$, $|\psi_2|^2 \approx |c_-|^2$, ψ_3, ψ_4 each has very small square modulus, and the third and fourth (error) terms correspond to dynamical states of α,β that, although each is mesoscopically determinate, are not such as to contain a record that the result was either a_+ or a_-. The interpretation of equation (31) is then as follows. After the observer-system β interacts with α (the apparatus-system), β will go into a mesoscopically determinate state in which it is almost certain to record either that the result was a_+ or that the result was a_-; the probability that it contains a record that the result was a_+, or a_-, is an extremely close approximation to the Born probability associated with the initial state of σ.

This still does not suffice to ensure that the record in β continues to faithfully correspond to the actual result of the measurement, as indicated by the dynamical state of α. Perhaps β comes to record the result a_+ although α indicated the result a_-? Once more, it is necessary to appeal to constraints on the evolution of dynamical states of subsystems to see why this is extremely unlikely. Two such constraints suffice.

The first is the Stability Condition of Healey (1989, 82), which implies that if it is true for the system representative of α at t that it recorded the value a_+ (a_-), then this is also true for the system representative of α at t'. Hence if α recorded a_- at t, then the system representative of α at t' is χ^{α}_{-}, and consequently the system representative of $\sigma \oplus \epsilon \oplus \beta$ at t' is $\psi_2 \oplus \chi^{\beta}_{-}$. The Minimal Meshing Condition (Healey 1989, 72) then implies that the system representative of β at t' is χ^{β}_{-}. Therefore β also records a_- at t'. (This application of the Stability Condition mirrors the discussion of the verifiability of measurement results on pages 90–92 of Healey [1989].) Now β will also interact with its environment, but as long as that interaction does not couple to the interaction of α with its environment, the constraint first applied in section 6 will suffice to secure the stability of β's record after t'. That constraint (applied here to the state of β) reads as follows: if Q_r, Q_s are quantities whose associated operators have identical eigenvalues and eigenspaces that are pairwise very close, and the value of Q_r is $s_+(s_-)$ at t, then if Q_s has a value a short time after t, it is almost certain that this value will also be $s_+(s_-)$. The quantities Q_r, Q_s on β may be thought of as approximate "record quantities," since the value of any such quantity on β provides a record in β of the result of the measurement by α on σ, and the above two dynamical constraints then guarantee that any such record is almost certain to be both faithful and stable.

Of course, there is still a sense in which even if one accepts this account of the generally faithful recording produced by an observer-system, a problem of explaining a conscious observer's experience of the measurement result remains. But this is now just the age-old problem of understanding how an observer's conscious experience could possibly result from purely physical changes in the state of his or her body. This problem may safely be left to philosophers of mind; if the present account of the functioning of a physical observer-system is correct, a detailed quantum-mechanical treatment of neurophysiological processes will be of no further help in solving that problem.

Notes

1. See Bub (1992, 1996, 1997), Dieks (1989a, b; 1994a, b), Healey (1989, 1995), Kochen (1985), van Fraassen (1973, 1991).

2. See Zurek (1981; 1982; 1991; 1993a, b), Unruh and Zurek (1989), Joos and Zeh (1985).

3. Including D'Espagnat (1990), Elby (1994), and various authors who replied to Zurek (1991) in *Physics Today* 46 (April 1993): 13–15, 81–82.

4. See Albert and Loewer (1990, 1993), Elby (1993).

5. See Bacciagaluppi and Hemmo (1996), Dickson (1994), Dieks (1994a, b), Elby (1994), Healey (1995).

6. See D'Espagnat (1976), Fehrs and Shimony (1974).

7. This argument implicitly relies on a couple of assumptions usually considered too obvious to be worth articulating, but in the present context these should be made explicit. One is that if quantity R is represented by self-adjoint operator $\hat{R}$ on the Hilbert space of α, then quantity R is represented by self-adjoint operator $\hat{I} \otimes \hat{R} \otimes \hat{I}$ on the Hilbert space of $\sigma \oplus \alpha \oplus \epsilon$. The other is that if a compound system is in a pure state, then the eigenvalue-eigenstate link is to be applied to that pure state in order to determine what observables on components of the compound system have sharp values.

8. Here I am indebted to Simon Saunders.

9. This is to treat the "branching" of conscious observer-processes in the way that Parfit (1984) argues we should treat hypothetical fission cases in his discussion of personal identity.

10. For technical details, see Healey (1989), chapter 2. The fact that the properties of a compound system are not determined by those of its subsystems introduces a kind of holism into the interpretation that proves important in accounting for violations of Bell inequalities.

11. Note that there are circumstances in which a system may be assigned as its quantum state a vector lying outside the subspace defining the system representative, and consequently fails to have a property assigned probability 1 by this quantum state. As noted previously, van Fraassen (1973) first introduced the term 'modal' to describe an interpretation according to which the quantum state specifies which properties a system *must* have (by virtue of assigning them probability 1), while the system *actually* has additional properties, in violation of the eigenvalue-eigenstate link. While my interactive interpretation does not satisfy this description, it rejects the eigenvalue-eigenstate link even more radically than interpretations that do satisfy it. Not only does it almost always ascribe dynamical properties to a system even though its quantum state does not assign these properties probability 1; there are also circumstances in which it fails to ascribe a dynamical property to a system, even though this system's quantum state does assign that property probability 1.

12. Much more needs to be said to defend this claim of empirical indistinguishability. This issue is taken up in the final section of the paper.

Bibliography

Albert, D., and B. Loewer. 1988. "Interpreting the Many-Worlds Interpretation." *Synthese* 77: 195–213.

———, and ———. 1990. "Wanted, Dead or Alive: Two Attempts to Solve Schrödinger's Paradox." In A. Fine, M. Forbes, and L. Wessels, eds., *PSA 1990, vol. 1*. Lansing: Philosophy of Science Association. 277–85.

———, and ———. 1993. "Non-Ideal Measurements." *Foundations of Physics Letters* 6: 297–305.

Bacciagaluppi, G. 1996. "Kochen-Specker Theorem in the Modal Interpretation of Quantum Mechanics." *International Journal of Theoretical Physics* 34: 1205–16.

Bacciagaluppi, G., and M. Hemmo. 1996. "Modal Interpretations, Decoherence and Measurements." *Studies in History and Philosophy of Modern Physics* 27B: 239–277.
———, and ———. 1994. "Making Sense of Approximate Decoherence." *PSA 1994, vol. 1:* 345–54. Lansing: Philosophy of Science Association.
Bub, J. 1992. "Quantum Mechanics without the Projection Postulate." *Foundations of Physics* 22: 737–54.
———. 1996. "Schrödinger's Cat and Other Entanglements of Quantum Mechanics." In J. Earman and J. Norton, eds., *The Cosmos of Science* (Pittsburgh: University of Pittsburgh Press).
———. 1997. *Interpreting the Quantum World.* Cambridge: Cambridge University Press.
D'Espagnat, B. 1976. *Conceptual Foundations of Quantum Mechanics,* 2d ed. Reading, Mass.: Benjamin.
———. 1990. "Toward a Separable 'Empirical Reality'?" *Foundations of Physics* 20: 1147–1172.
DeWitt, B. 1970. "Quantum Mechanics and Reality." *Physics Today* 23 (September): 30–35.
Dickson, M. 1994. "Wavefunction Tails in the Modal Interpretation." *PSA 1994, vol. 1:* 366–76. Lansing: Philosophy of Science Association.
Dieks, D. 1989a. "Quantum Mechanics without the Projection Postulate and Its Realistic Interpretation." *Foundations of Physics* 19: 1397–1423.
———. 1989b. "Resolution of the Measurement Problem through Decoherence of the Quantum State." *Physics Letters* A142: 439–446.
———. 1994a. "The Modal Interpretation of Quantum Mechanics Measurements, and Macroscopic Behavior." *Physical Review* A49: 2290–2300.
———. 1994b. "Objectification, Measurement, and Classical Limit According to the Modal Interpretation of Quantum Mechanics." In P. Busch, P. Lahti, and P. Mittelstaedt, eds., *Proceedings of the Symposium on the Foundations of Modern Physics*, Cologne 1993. Singapore: World Scientific.
Elby, A. 1993. "Why 'Modal' Interpretations of Quantum Mechanics Don't Solve the Measurement Problem." *Foundations of Physics Letters* 6: 5–19.
———. 1994. "The 'Decoherence' Approach to the Measurement Problem in Quantum Mechanics." *PSA 1994, vol. 1*: 355–65. Lansing: Philosophy of Science Association.
Everett, H., III. 1957. "Relative State Formulation of Quantum Mechanics." *Reviews of Modern Physics* 29: 454–62.
Fehrs, M., and A. Shimony. 1974. "Approximate Measurement in Quantum Mechanics II." *Physical Review* D9: 2321–23.
Gell-Mann, M., and J. Hartle. 1990. "Quantum Mechanics in the Light of Quantum Cosmology." In W. Zurek, ed., *Complexity, Entropy, and the Physics of Information*, 425–459. New York: Addison Wesley.
Ghirardi, G., A. Rimini, and T. Weber. 1986. "Unified Dynamics for Microscopic and Macroscopic Systems." *Physical Review* D34: 470–91.
Healey, R. 1984. "On Explaining Experiences of a Quantum World." *PSA 1984*, vol. 1: 56–69. Lansing: Philosophy of Science Association.
———. 1989. *The Philosophy of Quantum Mechanics: An Interactive Interpretation.* Cambridge: Cambridge University Press.
———. 1993a. "Why Error-Prone Quantum Measurements Have Outcomes." *Foundations of Physics Letters* 6: 37–54.
———. 1993b. "Measurement and Quantum Indeterminateness." *Foundations of Physics Letters* 6: 307–16.
———. 1995. "Dissipating the Quantum Measurement Problem." *Topoi* 14: 1–11.
Joos, E., and H. D. Zeh. 1985. "The Emergence of Classical Properties through Interaction with the Environment." *Zeitschrift für Physik* B59: 223–243.

Kochen, S. 1985. "A New Interpretation of Quantum Mechanics." In P. Lahti and P. Mittelstaedt, eds., *Symposium on the Foundations of Modern Physics 1985*, 151–69. Singapore: World Scientific.
Parfit, D. 1984. *Reasons and Persons*. Oxford: Clarendon Press.
Unruh, W. G., and W. H. Zurek. 1989. "Reduction of a Wave-packet in Quantum Brownian Motion." *Physical Review* D40: 1071–94.
Van Fraassen, B. 1973. "A Semantic Analysis of Quantum Logic." In C. Hooker, ed., *Contemporary Research in the Foundations and Philosophy of Quantum Theory*, 80–113. Dordrecht: Reidel.
———. 1991. *Quantum Mechanics: An Empiricist View*. Oxford: Clarendon Press.
Wigner, E. 1963. "The Problem of Measurement." *American Journal of Physics* 31: 6–15.
Zurek, W. H. 1981. "Pointer Basis of Quantum Apparatus: Into What Mixture Does the Wave Packet Collapse?" *Physical Review* D24: 1516–25.
———. 1982. "Environment-induced Superselection Rules." *Physical Review* D26: 1862–80.
———. 1991. "Decoherence and the Transition from Quantum to Classical." *Physics Today* 44: 36–44.
———. 1993a. "Preferred States, Predictability, Classicality, and the Environment-induced Decoherence." *Progress of Theoretical Physics* 89: 281–312.
———. 1993b. "Negotiating the Tricky Border between Quantum and Classical." *Physics Today* 46: 13–15, 81–90.

Andrew Elby

Interpreting the Existential Interpretation

Since its introduction several years ago, Wojciech Zurek's "existential interpretation" (Zurek 1993a, b) has incited passionate debate among physicists and philosophers. Most philosophers remain puzzled about his interpretation. But at the conference on "Decoherence and Modal Interpretations" in Minneapolis in 1995, we held a discussion in which Zurek clarified his interpretational stance. He was responding to my "commentary" about his paper, in which I displayed and discussed a bunch of interpretational "boxes." After arguing that Zurek's interpretation must fit into one of the boxes, I asked him which one. Zurek's answer, and the ensuing discussion, helped to clear up some of the confusion surrounding the existential interpretation.

In this chapter, I lay out and comment on these Minneapolis discussions, as informed by subsequent e-mail correspondence with Zurek, who, it turns out, views his interpretation not as a metaphysically complete description of the universe, but as a schema that focuses on the relationship between the observer-participator within that universe and the "facts" that are "quenched out" from the universal wave-function by the process of decoherence. He remains largely agnostic about the kinds of metaphysical issues over which philosophers agonize. Unfortunately, he or some of his supporters may verge on advocating an untenable variety of "open-mindedness" about interpretational matters. And some of Zurek's supporters (although not Zurek himself) may be overly optimistic about the extent to which further formal developments in decoherence theory can shed light on certain kinds of metaphysical questions.

1. The Box Argument

In this section, I present my argument that the existential interpretation must fit into a well-defined interpretational "box." The existential interpretation invokes decoherence to select a preferred set of physical quantities. Although this set may or may not constitute a basis in Hilbert space, let me call it a "basis" for brevity. For my purposes, the specifics of Zurek's (1993) basis-selection rule don't matter. But here they are, briefly. To find the "special" physical quantities associated with a system, examine the interaction Hamiltonian, $\mathbf{H}_{\text{int}}$, describing the system's interplay with its environment. Of all the observables associated with the system, find the ones that commute with $\mathbf{H}_{\text{int}}$. If $[\mathbf{H}_{\text{int}}, \mathbf{A}] = 0$, where $\mathbf{A}$ "inhabits" the Hilbert space associated with the system alone, then the eigenvectors of $\mathbf{A}$ are (part of) the preferred basis for that system. In other words, the system possesses a definite value for observable A, in some (perhaps observer-relative) sense.

Zurek's technical rules are clear. For concreteness, let's say the system under consideration is a measuring apparatus, for which the relevant interaction Hamiltonian selects a pointer-reading operator $\mathbf{A}$ with two eigenvectors, $|A = a_1\rangle$ and $|A = a_2\rangle$. In many cases, decoherence ensures that the final quantum state of the apparatus/environment is

$$\Psi = c_1 \,|A = a_1\rangle \otimes |E_1\rangle + c_2 \,|A = a_2\rangle \otimes |E_2\rangle,$$

where the environmental states $|E_1\rangle$ and $|E_2\rangle$ are very nearly orthogonal. Despite the fact that Ψ isn't an eigenstate of $\mathbf{A} \otimes \mathbf{I}$, the pointer has a definite pointer reading in some objective or subjective sense, either a_1 or a_2 (or perhaps both, à la Everett). Here's my central question: In what sense is the pointer reading definite in the existential interpretation?

The answer, I claim, *must* fall into exactly one of the following boxes, or else the interpretation becomes incoherent. My first wedge divides hidden-variable and modal interpretations from relative state interpretations. According to Zurek, the pointer reading is definite in some (perhaps observer-relative) sense; it is *not* the case that *neither* a_1 nor a_2 becomes actualized. Therefore, either exactly one of the possible pointer readings is actualized (occurrent), or else they are both actualized (occurrent) in some sense. In other words, either (i) one of the pointer readings actually *obtains* while the other does not; or (ii) both pointer readings obtain. Under option (ii), of course, a given observer's mind perceives exactly one pointer reading.

Option (i) corresponds to Bohm-like and modal-like interpretations, according to which exactly one branch of the above superposition corresponds to an instantiated property. The other branch, though dynamically relevant, does not correspond to a property of the system.

Option (ii) corresponds to Everett-like interpretations. Both branches of the superposition are "actualized" in some sense, either objective or subjective. Options (i) and (ii) exhaust the logical possibilities about what it means for the pointer reading to be definite.

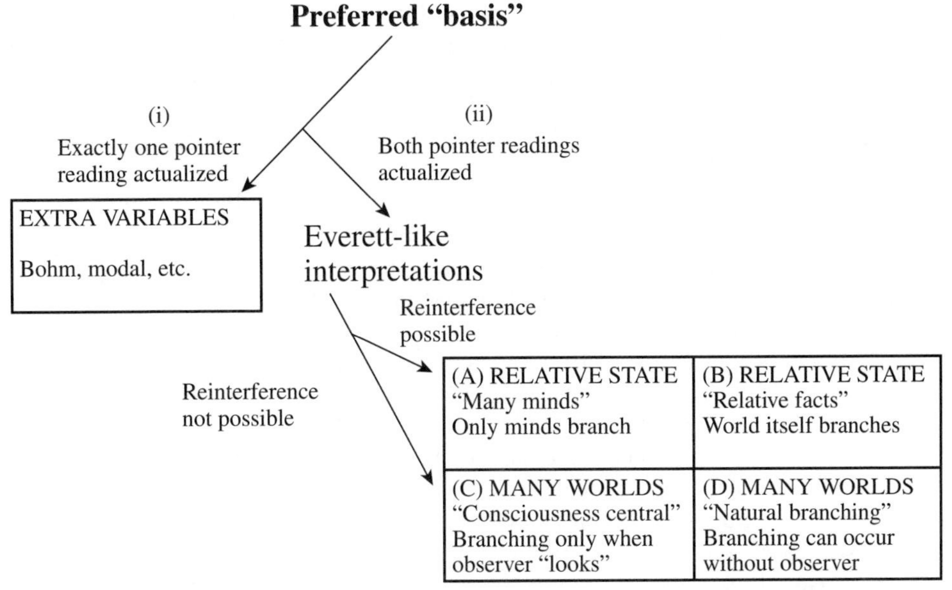

Figure 1. Boxes into which a preferred-basis interpretation must fit.

We can partition box (ii) into logically exhaustive subboxes. If both pointer readings are actualized in some sense, then they can either be actualized in the same world or in different worlds. To formalize this wedge, consider the reinterference question: Is it possible, in principle, for the two different branches of the superposition to interfere with each other? By "many-worlds" interpretations, I mean option (ii) interpretations that answer "no" to this question; by one-world "relative state" interpretations, I mean option (ii) interpretations that answer "yes."[1]

Within the one-world view, we can erect another partition. The actualization of the two branches of the superposition takes place either (A) solely in the mind(s) of the observer, not in the physical world itself; or (B) in the physical world itself. Box (B) doesn't preclude the branching of minds, but under (B), any such mind-branching occurs in addition to, not in place of, a relative-state branching of the pointer itself. Boxes (A) and (B) exhaust the possibilities for one-world relative state interpretations.

In box (A), the world itself stays unbranched, completely described by the superposition. But the observer's mind "branches," and each branch perceives a definite pointer reading. This is the box, therefore, into which many-minds interpretations fit. See Albert and Loewer (1988) for an example. Clearly, box (A) interpretations rest on an aggressively dualistic metaphysics.

Under box (B), there's just one world, but many "actualizations" of the pointer reading. As Simon Saunders (1995) argues, this view can be rendered coherent only if we radically change our metaphysics. We must view a property not as something that gets instantiated in the world, but as something that gets instantiated *relative* to other instantiated properties. For instance, it's not the case that I exist. All we can say is that I exist *relative to* Bill Clinton's being president, Saturn's having rings, and all the other physical facts belonging to "my" branch of the superposition. I may find this disturbing—after all, I would like to view my existence as something absolutely true about the world. But box (B) allows no such absolutism. All facts and properties are relative.

Similarly, we can partition the many-worlds view into two exhaustive possibilities. Either (C) world splitting occurs *only* when an observer "looks," or else (D) world splitting can occur at other times. In box (C), the observer's mind has a special influence; we can "split" worlds just by looking! In box (D), the observer plays a less crucial role. But advocates of (D) must specify under what circumstances worlds split. Similarly, advocates of box (C) must spell out who counts as an observer and what counts as "looking."

At the risk of beating a Schrödinger's horse, I can't overemphasize that boxes (A) through (D) are a complete orthogonal set within option (ii). An Everett-like interpretation must fall into exactly one of those boxes—there's no other logical possibility. And as we have seen, Zurek's existential interpretation must fall either into the Everett-like box or into the "extra variables" box.

At the conference in Minneapolis, after laying this out, I then asked Zurek into which box his interpretation fits.

2. Zurek's Response

As Zurek expressed during the ensuing discussion at the conference, and also in subsequent e-mail correspondence with me, he resists being forced into a box. He expresses a slight preference for (B), "relative facts," but also claims that the boundaries of the boxes might "stretch." Zurek says that, ultimately, he's agnostic about how "reality" fits into the whole scheme. He prefers to focus solely on the perceptions of the observer. He expresses the view that further formal work will clarify the meaning of the existential interpretation, though probably only marginally. Language itself, he argues, circumscribes the range of interpretational options we humans can formulate.

Zurek's clarification of his stance helps us to clear up much of the frantic confusion surrounding his interpretation. He does not view the existential interpretation as a polished "final word" about reality, and resists getting pinned down in philosophical discussions about these issues.

3. What Is Open-mindedness?

When David Albert and I argued that Zurek's interpretation must fit into one of my boxes, some participants argued forcefully that as philosophers, we must remain open-minded about these difficult interpretational issues. I will now identify what kind of open-mindedness we should embrace and what kind we should reject.

The discussion at the conference conflated two kinds of open-mindedness. On the one hand, we can accept that Zurek's interpretation must fit into one of my boxes, and we remain open-minded about which box is best. On the other hand, we can remain open-minded about whether an interpretation must fit into a box.

To my knowledge, everyone agrees that we should remain open-minded in the first sense, carefully weighing new arguments about the merits of relative facts versus many minds, or about new modal interpretations, for example. But open-mindedness in the second sense amounts to abandoning the quest for consistency and coherence. By construction, my boxes capture all possible senses in which an observable can be "definite" in a no-collapse interpretation. Certainly, if I've committed an error of logic, then correct my boxes accordingly. And if you want to repartition the space of all possible Everett-like interpretations, that's fine too; provided your boxes are disjoint and logically exhaustive, an Everett interpretation must fit into one of your boxes as well as into one of mine. But we *cannot* remain open-minded about whether an interpretation must fit into exactly one of the boxes, if the boxes are indeed disjoint and exhaustive.

This is not to say that we cannot improve our understanding of the philosophical implications associated with each box. Philosophers can contribute to the debate by spelling out the advantages and disadvantages of interpretations that fall within the different boxes. We can formulate new interpretations inside each box. But it's also our role to set limits, by exhaustively delineating a set of distinct alternatives (when possible). If the boxes really are distinct and exhaustive, then we can't stretch the boundaries, and there's little point in waiting for new alternatives to arise. All we can do is try to understand the alternatives better, and chop up the space of alternatives in different ways.

4. The Role of Formal Developments in Interpretation

In this section, I discuss the effect of new formal developments on the interpretation of quantum mechanics. Since this touches on a deep and messy issue in the history and philosophy of science, I won't generalize. Instead, I'll build on the previous section to spell out what formal developments in decoherence can and cannot accomplish, with respect to foundations of quantum mechanics.

Some decoherence theorists express the hope that by better understanding decoherence (and hence, the whole quantum formalism), we can discover new interpretive options, and make better choices between them. I now argue that the only sense in which further formal developments can shed light on these interpretive issues is a narrow technical one. A God's-eye understanding of decoherence could not, even in principle, supply us with arguments about which box is best.

Decoherence can tell us which observables get "picked out" by the environment, according to various basis-selection rules. Zurek's requirement that $[\mathbf{H}_{\text{int}}, \mathbf{A}] = 0$ is one such rule. The biorthogonal decompositions invoked by some modal interpretations is another. Further formal work can increase our understanding of which observables get picked out in which situations by which rules. For instance, Zurek and his colleagues may one day discover that the observable selected by the biorthogonal decomposition rule isn't always close to the position observable, even for macroscopic objects whose positions we perceive as definite. If this happens, then some modal theorists will need to revise their basis-selection rule, perhaps using Zurek's rule instead of biorthogonal decompositions. "Decoherence theory" can thus contribute greatly to the foundations of quantum theory—indeed, it has already done so. But in this context, decoherence theory can accomplish nothing more than to help interpreters pick an adequate basis-selection rule. These formal developments cannot, even in principle, help us decide how to *interpret* the preferred basis.

I am not arguing that formal developments don't fuel foundational progress. They certainly do. Rather, I'm making a specific argument about this specific case study. Decoherence theory can help interpretations that rely on a preferred basis to explore the advantages and disadvantages of different basis-selection rules. But it can't help us to decide more purely metaphysical questions, such as which Everett-like box is best. We can't fool ourselves into thinking that further formal developments in decoherence theory will push us toward one box, or will stretch the boundaries so as to open up new options. Although Zurek himself doesn't fall into this trap, other decoherence theorists seem to feel that a better understanding of decoherence will solve all our problems. It won't.

5. Conclusion

I have presented Zurek's clarification of his interpretational stance. The existential interpretation, rather than giving us a metaphysically complete picture of reality, focuses on the relationship between the observer and the "facts" perceived. When presented with a catalog of interpretations that differ only in their metaphysical details, Zurek chooses to remain agnostic. For this reason, philosophers would be well advised to stop pressing Zurek to clarify his interpretation.

Exploring the nature of this agnosticism, I argued that formal developments in decoherence theory cannot help sort out all these metaphysical problems. These formal developments can only help interpreters choose a good basis-selection rule. I also teased apart two senses of open-mindedness, and insisted that we not remain open-minded about whether an interpretation should be forced into a box, when the boxes are disjoint and cover all logical possibilities.

Notes

1. As Martin Jones (personal communication) points out, we can concoct a many-worlds interpretation in which reinterference is allowed. Suppose that there are, in fact, many worlds. In each world, the dynamical state is y, the wave-function predicted by Schrödinger's equation. Crucially, each world "gets" the whole wave-function, not a single branch of the wave-function. Within each world, the wave-function gets interpreted as it does in modal interpretations; although the whole wave-function is dynamically relevant, the "property state" of a given subsystem corresponds to just one branch.

In my box scheme, this "deWitt-modal" interpretation is considered to be a one-world relative state interpretation. My wedge, though useful for dividing Everett-like interpretations into logically exhaustive possibilities, does not fully capture the intuitive split between one-world and many-worlds viewpoints. Note, however, that the actual one-world and many-worlds views advocated in the literature are indeed split by my wedge. After all, in the deWitt-modal interpretation, the "extra" worlds do not do any philosophical work, since each world "gets" the whole wave-function. Therefore, an advocate of the deWitt-modal interpretation has no incentive not to delete all the superfluous metaphysics (that is, the extra worlds) and become a "regular" one-world modal theorist. An advocate of one-world relative state interpretations in which the Schrödinger equation is universally valid cannot coherently claim that reinterference is impossible.

References

Albert, D., and B. Loewer. 1988. "Interpreting the Many-Worlds Interpretation." *Synthese* 77: 195–213.

Bohm, D., B. Hiley, and P. Kaloyerou. 1987. "An Ontological Basis for the Quantum Theory." *Physics Reports* 144: 321–75.
Saunders, S. 1995. "Time, Quantum Mechanics, and Decoherence." *Synthese* 102: 235–66.
Zurek, W. 1993a. "Preferred States, Predictability, Classicality, and the Environment-Induced Decoherence." *Progress in Theoretical Physics* 89: 281–312.
———. 1993b. "Negotiating the Tricky Border between Quantum and Classical: Zurek Replies." *Physics Today* 46 (4): 84–90.

Guido Bacciagaluppi and Meir Hemmo

State Preparation in the Modal Interpretation

1. Introduction

Most versions of the modal interpretation of quantum mechanics are based on van Fraassen's (1973) intuition that the uncollapsed reduced state of a quantum-mechanical system, which is generally a mixed state, describes in the first place not the actual properties but the possible properties of the system.

According to van Fraassen (1973; 1991, chapter 9), the projections that appear in any decomposition of the reduced state of the system all represent possible properties. In most of the later versions of the modal interpretation, a specific decomposition of the reduced state is privileged, namely, the spectral (orthogonal, diagonal) decomposition. These versions are notably the ones by Kochen (1985), Krips (1987), Healey (1989), Dieks (1989, 1994), Clifton (1995), and the generalization by Vermaas and Dieks (1995). These versions have many important differences, but they all share a privileged role assigned to the spectral decomposition of the state. A different framework that overlaps with the above modal interpretations has been proposed by Bub (1992, 1997).[1]

The central question is whether this privileged role assigned to the spectral decomposition is physically justified, in the sense of empirical adequacy. This question is independent of the more abstract justifications recently given by Clifton (1995) and by Dieks (1995). The problem of the empirical adequacy of the spectral decomposition was forcefully raised by Albert and Loewer (1990) in their criticism of the modal interpretation based on the analysis of nonideal measurements (see also Ruetsche [1995]). Quite generally, Albert and Loewer doubt that in the case of macroscopic systems, the spectral decomposition will always correspond to the properties that we expect to be definite, given our everyday experience.

In a previous paper (Bacciagaluppi and Hemmo 1996), we argued that the case of macroscopic systems needs to be examined in the framework of the theory of decoherence (see, for example, Zurek [1993] and references therein), because a macroscopic system is continuously in interaction with its environment. We have examined in detail the situation in the models of measurement in which the apparatus is considered as a finite-dimensional system, in the sense that the different pointer states form a basis in the Hilbert space of the apparatus, and where the environment couples to these pointer states. In these models, apart from cases of extreme degeneracy, the spectral decomposition of the state of the apparatus will be in terms of states close in scalar product to the ideal pointer states. However, macroscopic systems are infinite-dimensional, or at least very high-dimensional systems, and the interaction with the environment will lead to extremely mixed (and degenerate) states of the macroscopic system. It is thus still doubtful whether in the case of macroscopic systems the spectral decomposition of the state is empirically adequate. Some strong arguments against its adequacy are given by Bacciagaluppi, Donald, and Vermaas (1995), Donald (1998), and Bacciagaluppi (1996b).

We shall set these questions aside here and examine a related but different question, namely, the question of state preparation. For the purposes of the discussion, we shall take the models of measurements in which the apparatus is a finite-dimensional system. Thus, we shall in fact assume that the spectral decomposition of the reduced state of all our pieces of apparatus is in terms of properties that are empirically adequate. Nevertheless, a problem of *state preparation* arises, that is, the problem of the relation between the properties of the measured system and the preparatory aspects of the measurement. In the second half of this chapter we shall examine the properties assigned to the system in the modal interpretation from the point of view of hidden-variables theories.

We shall be working in the full formalism of the Vermaas and Dieks (1995) version. This version assigns joint probability distributions for the possessed properties of subsystems $\alpha, \beta, \ldots \omega$ in any given factorization of a composite system $\alpha \otimes \beta \otimes \ldots \otimes \omega$, which need not be in a pure state. In this generalization, the correlations between the possessed properties of the subsystems need not be one-to-one. Earlier versions of the modal interpretation are based on the biorthogonal (Schmidt) decomposition theorem, and assign joint probabilities only to pairs of systems when their composite is in a pure state; in this case, the correlations are always perfect. Vermaas and Dieks propose that joint probabilities for possessed properties (at a given time) be given by the formula

$$Tr(\rho^{\alpha\beta\ldots\omega}\, P^{\alpha} \otimes P^{\beta} \otimes \cdots \otimes P^{\omega}), \tag{1}$$

where $P^{\alpha}, P^{\beta} \ldots P^{\omega}$ are, respectively, the definite properties of the systems $\alpha, \beta, \ldots \omega$, as given by the spectral resolutions of the systems' reduced states, and $\rho^{\alpha\beta\ldots\omega}$ is the (generally mixed) state of the composite system $\alpha \otimes \beta \otimes \ldots \otimes \omega$.

Our discussion will also make use of some results about the dynamical evolution of possessed properties. We shall use the notion of *transition probabilities* between possessed properties at different times (Vermaas 1996). For a detailed discussion of dynamics in the modal interpretation, we refer to Bacciagaluppi and Dickson (1997). The use of transition probabilities will highlight the probabilistic dependence of measurement outcomes on possessed properties of other systems. This, in turn, will lead us to discuss the picture of the modal interpretation (at least in the Vermaas and Dieks [1995] version) as a stochastic hidden-variables theory.

The paper is structured as follows. In section 2, we analyze the problem of state preparation in detail. In section 3, we suggest that, under the stated assumption of definiteness of pointer readings, and a further assumption of stability, the modal interpretation can solve this problem by an analysis of preparation and registration devices. The discussion of the modal interpretation as a stochastic hidden-variables theory follows in section 4. We conclude in section 5 with a discussion of the theory's nonlocality and of Healey's (1989) views on holism in the modal interpretation.

2. The Problem of State Preparation

In standard quantum mechanics, one often assigns to quantum systems pure states, or mixed states that are proper mixtures. In particular, after a measurement, it is assumed that the quantum state of the measured system has undergone a collapse, and the system is considered to have been prepared in the corresponding state. If the measurement is selective, this state can be a pure state. If it is nonselective, it is always a proper mixture. Further predictions are based on the assignment of these states and are amply confirmed by experimental results.

Vermaas and Dieks (1995) emphasize that, strictly speaking, there are no proper mixtures in the modal interpretation, because it is a no-collapse interpretation. In the standard theory, however, these states play an important role, which needs to be recovered by the modal interpretation. Standard quantum mechanics makes a distinction between a proper and an improper mixture corresponding to the same ρ (or equivalently, between an inhomogeneous and a homogeneous ensemble), namely when there is the possibility of making a *selection* of a subensemble. Standard quantum mechanics does make predictions for these cases, in particular predictions about outcomes of further measurements.

In a nutshell, the problem of state preparation is the following. The modal interpretation aims at explaining the predictions of standard quantum theory by assigning properties to (microscopic or macroscopic) systems. In ideal situations, these properties are the same as those assigned to the systems by a collapse interpretation. In particular, when microscopic systems are ideally prepared in certain states, then according to the modal interpretation they possess the corresponding properties, even though it is assumed that there is no collapse. It thus seems that the modal interpretation can explain the further behavior of those systems by virtue of their possessed properties alone. In general, however, the quantum-mechanical states assigned to microscopic systems in the standard theory will not correspond to the properties assigned to the same systems by the modal interpretation. If this is the case, then the question arises whether and how in the modal interpretation possessed properties can still be taken to explain the further predictions of quantum mechanics. In addressing this problem, the modal interpretation will have to recover the notion of a selection of a subensemble in a selective measurement, as well as the Born rule of standard quantum mechanics.

We can always prepare a system in some state by means of a suitable interaction with a measuring apparatus. In the following, we shall understand the latter to be a finite-dimensional system (as mentioned above), with a privileged orthonormal basis $\{|\psi_i\rangle\}$ of pointer reading states, including a "ready state" $|\psi_0\rangle$. We shall assume that the initial state of such a measurement interaction is a product state of a (generally pure) state of the system, say $|\phi\rangle$, and the ready state of the apparatus $|\psi_0\rangle$.

This preparatory aspect of a measurement interaction is as follows. At the end of the interaction (but before any supposed collapse), the state of the combined system consisting of the quantum system and the preparation device has the general form

$$\sum_{i,j} \lambda_{ij} \, |\phi_i\rangle \otimes |\psi_j\rangle, \tag{2}$$

where the $|\phi_i\rangle$ form some basis in the Hilbert space of the system, and where the $|\psi_j\rangle$ are the pointer states of the apparatus (in fact, any state of the composite has this form). By performing first the sum over the index i, we can write the state as

$$\sum_{j} \mu_j |\phi_j^*\rangle \otimes |\psi_j\rangle. \tag{3}$$

The $|\phi_j^*\rangle$ here are the relative states of the system with respect to the pointer states $|\psi_j\rangle$, and in general they need not be orthogonal. In the special case in which $\lambda_{ij} = \mu_j \delta_{ij}$, the measurement is an ideal measurement of an observable A with eigenstates $|\phi_j\rangle$ (and in particular $|\phi_j^*\rangle = |\phi_j\rangle$ for all j).

The reduced state of the quantum system is given by the density matrix

$$\sum_j |\mu_j|^2 |\phi_j^*\rangle\langle\phi_j^*|. \qquad (4)$$

At this stage, this decomposition of the density matrix is an *improper* mixture, that is, it does not admit an ignorance interpretation. Yet the collapse postulate states that the state of the quantum system is transformed to one of the states $|\phi_j^*\rangle$ with probability $|\mu_j|^2$.[2] In other words, it is postulated that the decomposition (4) of the reduced state is (or becomes) a *proper* mixture.

The difference between a selective and a nonselective measurement is that in a selective measurement, one eliminates all or part of the ignorance by selecting a subensemble from the original ensemble, according to the outcome of the measurement. For instance, one can select those systems for which the outcome of the measurement is given by the pointer reading $|\psi_j\rangle$. Each of these systems will be in the state $|\phi_j^*\rangle$. Predictions for the outcomes of future measurements on a system in that subensemble (using the Born rule) are then based on the ascription of the state $|\phi_j^*\rangle$ to the system. If the $|\phi_j^*\rangle$ are nonorthogonal, the decomposition of the system's reduced state given by (4) which determines the properties of the system according to standard quantum mechanics is *not* orthogonal. Often, one considers only ideal measurements, for which $|\phi_j^*\rangle = |\phi_j\rangle$, as "preparatory." Any measurement, however, will leave the system in some state, and further predictions have to be based on that state, irrespective of whether it is an eigenstate of the measured observable.

The properties assigned to the quantum system by the modal interpretation depend on the uncollapsed reduced state (4) of the system. The modal interpretation postulates that one should take the spectral decomposition of (4), and interpret *this* decomposition of the state as if it were a proper mixture. The spectral decomposition of the state of the system has the form

$$\sum_j |\hat{\mu}_j|^2 |\hat{\phi}_j\rangle\langle\hat{\phi}_j|, \qquad (5)$$

where the $|\hat{\phi}_j\rangle$ will be orthogonal and in general different from the $|\phi_j^*\rangle$, and the $|\hat{\mu}_j|^2$ will in general be different from the $|\mu_j|^2$.

One can now draw two conclusions: first, that the properties assigned to the quantum system by the modal interpretation are generally *different* from the ones assigned by standard quantum mechanics; and second, that the subensembles selected in a selective measurement according to the measurement outcome will be *inhomogeneous* with respect to the properties assigned by the modal interpretation, that is, they will contain systems which, though they have the same collapsed state $|\phi_j^*\rangle$, have different possessed properties $|\hat{\phi}_j\rangle$. This is obvious, because the $|\hat{\mu}_j|^2$ are different from the $|\mu_j|^2$, so the subensembles characterized by different $|\hat{\phi}_j\rangle$ will be even different in size from the subensembles characterized by different $|\phi_j^*\rangle$.

In the special case of ideal measurements, the decomposition selected by the collapse postulate, namely

$$\sum_j |\mu_j|^2 |\phi_j\rangle\langle\phi_j|, \tag{6}$$

is already orthogonal. Thus, the modal interpretation assigns to the quantum system in this ideal case exactly the same properties as standard quantum theory, and with the same probabilities. In this case, it seems that all predictions of quantum mechanics for the outcomes of future measurements might be explained by the modal interpretation in terms of the possessed properties of the system. In general, however, this strategy will not work.

Suppose that after a selective measurement the system collapses to the state $|\phi_1^*\rangle$, which happens to be an eigenstate of spin in the x direction, and suppose that the reduced state of the system is diagonal in the basis $|\hat{\phi}_j\rangle$ of eigenstates of spin in the x' direction. Then the system will possess a definite spin in the x' direction, with a certain probability distribution for its value, but it will not possess any definite spin in the x direction. Suppose now we measure the spin in the direction x. According to standard quantum mechanics, we will obtain *with probability 1* an outcome corresponding to $|\phi_1^*\rangle$. But the system has no property that corresponds to spin x. The modal interpretation cannot explain this prediction with probability 1 by saying that the system has, indeed, a corresponding property. But then, how can the modal interpretation explain this prediction at all?

Albert and Loewer's (1990) original criticism focused on the properties of the apparatus. They showed that in the case of nonideal measurements, if system and apparatus are considered in isolation, the properties of the apparatus do not correspond to the ones predicted by standard quantum mechanics. In their criticism, the properties of the measured system were unimportant, because although the modal interpretation and the standard theory assign different properties to the measured system, it is the properties of the apparatus that in our experience seem to correspond to the ones predicted by standard quantum mechanics. On the other hand, the properties of the system are *prima facie* expected to determine the outcomes of *further* measurements. Thus, it seems that again, if the measurement is nonideal, the modal interpretation will have difficulty in recovering well-confirmed predictions of the standard theory. Further, while the assignment of properties to the apparatus depends on the inclusion of the environment (as done in Bacciagaluppi and Hemmo's [1996] treatment of Albert and Loewer's criticism), the properties assigned to the system do not depend on whether one does or does not consider the environment, because the system does not (necessarily) interact with the environment. Thus, this new problem—while in many ways analogous to the one raised by Albert and Loewer—requires a new kind of solution.

3. Preparation and Registration Devices

As the preceding discussion shows, in order to deal satisfactorily with the problem of state preparation, the modal interpretation must satisfy two *desiderata*:

(A) The modal interpretation ought to give a characterization in terms of possessed properties of the subensembles that are selected in the standard theory.

(B) The modal interpretation ought to recover the probabilistic predictions of the standard theory for the outcomes of further measurements (the Born rule).

Let us first address *desideratum* (A). The answer is simple. It is true that in the standard theory, one can characterize the subensembles selected in a measurement in terms of the states of the systems in the subensemble. This is not the criterion on which the selection is based in the first place, however. Rather, the selection is based on the *pointer readings* $|\psi_j\rangle$ of the preparation device. Here, we assume that the finite-dimensional analysis of Bacciagaluppi and Hemmo (1996) applies, and thus that the definite properties $|\hat{\psi}_j\rangle$ of the apparatus (given by the spectral resolution of its state) are close to the ideal pointer states $|\psi_j\rangle$. If we assume that these properties $|\hat{\psi}_j\rangle$ explain our experience of a definite reading of the pointer (as discussed in Bacciagaluppi and Hemmo [1996] and in Bacciagaluppi, Elby, and Hemmo [1996]), then we can assert that when we select a subensemble of measured systems according to the measurement outcomes, we operate this selection on the basis of the properties $|\hat{\psi}_j\rangle$ of the apparatus. Therefore, it is indeed possible to characterize the subensembles selected in the standard theory in terms of some possessed properties; however, these possessed properties are not properties of the *system*, but properties of the *preparation device*.

Thus, *desideratum* (A) can be fulfilled. We have a characterization of the subensembles that is internal to the modal interpretation. We can now see very clearly, using the Vermaas-Dieks correlation rule (1), that, as mentioned above, the subensembles selected on the basis of the pointer readings are not homogeneous with respect to the possessed properties of the system. In the case of a nonideal measurement, the states $|\phi_j^*\rangle$ need not coincide, not even approximately, with the possessed properties $|\hat{\phi}_j\rangle$. We thus have

$$|\phi_j^*\rangle = \sum_i \mu_i^j |\hat{\phi}_i\rangle, \tag{7}$$

with some coefficients μ_i^j that need not vanish for $i \neq j$. The final state (3) (including the environment) will have the form

$$\sum_{i,j} \mu_j \mu_i^j |\hat{\phi}_i\rangle \otimes |\psi_j\rangle \otimes |E_j\rangle, \tag{8}$$

where the $|E_j\rangle$ are approximately orthogonal, and thus the possessed properties $|\hat{\psi}_j\rangle$ of the apparatus are close to the states $|\psi_j\rangle$ (again, except in rare cases of near-degeneracy). By the Vermaas-Dieks correlation rule (1) then, the joint probability for the occurrence of $|\hat{\phi}_i\rangle$ and $|\hat{\psi}_j\rangle$ is approximately equal to $|\mu_j \mu_i^j|^2$. This means that the properties of the system and the properties of the apparatus are not perfectly correlated, and a subensemble selected on the basis of the properties of the apparatus (respectively, the system) will not be homogeneous with respect to the properties of the system (respectively, the apparatus).

We now turn to *desideratum* (B). The Born rule of standard quantum theory yields probabilistic predictions for outcomes of measurements, conditional on the state of the measured system, which one assumes to have been prepared by a procedure like the above. The state of the system then labels a particular subensemble of systems that have been selected at the end of the preparation. We have seen that in the modal interpretation, such a subensemble is *not* characterized by some property of the systems in the subensemble but rather by some property of the preparation device. In order to recover the probabilistic predictions of the standard theory, in particular the Born rule, we thus *cannot* conditionalize on the possessed property of the system. Rather, we shall have to conditionalize on the possessed property of the preparation device.

Let us introduce two pieces of apparatus, a first measuring apparatus M_1, which functions as a preparation device, and a second measuring apparatus M_2, which functions as a registration device (both will be subject to decoherence). We need to calculate the correlations between the possessed properties of M_1 at the time t_1 when the preparation is completed, and of M_2 at the time t_2, when the registration is completed. On analogy with (3), we take the (uncollapsed) state at t_1 (after the preparation) to be

$$|\Psi_1\rangle = \sum_i \mu_i |\phi_i^*\rangle \otimes |\psi_i^1\rangle \otimes |\psi_0^2\rangle \otimes |E_{i0}\rangle, \tag{9}$$

where the environment vectors are approximately orthogonal, so that except in extremely degenerate cases, the reduced state of M_1 is diagonal in a basis, call it $|\hat{\psi}_i^1(t_1)\rangle$, that is close to the basis $|\psi_i^1\rangle$.

We now assume that the registration is a measurement of an observable $A = \Sigma_j \alpha_j |\eta_j\rangle\langle\eta_j|$. This is the case to which the Born rule is applicable in its usual form.[3] One can write each $|\phi_i^*\rangle$ as a superposition of the eigenstates of A:

$$|\phi_i^*\rangle = \sum_j \nu_j^i |\eta_j\rangle. \tag{10}$$

In order to simplify the calculations, we also assume that the measurement of A is perfect, in the sense that the interaction perfectly couples the eigenstates of A to the pointer states: for any coefficients ν_j,

$$\sum_j \nu_j |\eta_j\rangle \otimes |\psi_0^2\rangle \mapsto \sum_j \nu_j\, |\eta_j^*\rangle \otimes |\psi_j^2\rangle \tag{11}$$

(notice that the $|\eta_j^*\rangle$ can be arbitrarily disturbed states, as in a measurement of the second kind). Finally, let us assume for the time being that the registration starts immediately after the preparation is completed; in other words, the final state of the preparation $|\Psi_1\rangle$ is also the initial state of the registration.

The final state at t_2, when the registration is completed, has then the form:

$$|\Psi_2\rangle = \sum_i \mu_i \sum_j \nu_j^i |\eta_j^*\rangle \otimes |\psi_i^1\rangle \otimes |\psi_j^2\rangle \otimes |E_{ij}\rangle, \tag{12}$$

where the environment vectors are approximately orthogonal in *both* indices i and j. This amounts to no reinterference of the branches in the evolution of the state (9). Thus, the definite properties $|\hat{\psi}_i^1(t_2)\rangle$ of the preparation device M_1 are still close to the eigenstates $|\psi_i^1\rangle$ of its pointer observable, and similarly, the definite properties $|\hat{\psi}_j^2(t_2)\rangle$ of the registration device M_2 are close to the eigenstates $|\psi_j^2\rangle$ of its pointer observable.[4]

In this case, by (10) and (12), the standard theory predicts that the probability of an outcome $|\psi_j^2\rangle$ for the registration, conditional on a preparation of the state $|\phi_i^*\rangle$, is equal to

$$|\nu_j^i|^2 = |\langle\phi_i^*|\eta_j\rangle|^2. \tag{13}$$

These are the Born probabilities. We now claim that the correlations between the properties of M_1 at t_1 and the properties of M_2 at t_2 are (approximately) the same as given by (13).

We need to calculate the conditional probability

$$p(\hat{\psi}_j^2(t_2)|\hat{\psi}_i^1(t_1)). \tag{14}$$

We shall now assume *stability* of the properties of M_1. That is, we shall assume that for all i,

$$p(\hat{\psi}_i^1(t_2)|\hat{\psi}_i^1(t_1)) \approx 1, \tag{15}$$

which also implies that for all i,

$$p(\hat{\psi}_i^1(t_1)|\hat{\psi}_i^1(t_2)) \approx 1. \tag{16}$$

Stability thus means that at t_2 there is a *faithful record* of the outcome of the preparation: if M_1 shows a certain reading at t_1, it will not only have *some* definite reading also at t_2, but it will have the *same* reading.

Thus we obtain:

$$p(\hat{\psi}_j^2(t_2)|\hat{\psi}_i^1(t_1)) \approx p(\hat{\psi}_j^2(t_2)|\hat{\psi}_i^1(t_2)) = \frac{p(\hat{\psi}_j^2(t_2)\,\&\,\hat{\psi}_i^1(t_2))}{p(\hat{\psi}_i^1(t_2))} \tag{17}$$

Now we can calculate these probabilities using the rules of the modal interpretation at the time t_2. In particular, by applying the Vermaas-Dieks correlation rule to (12), we get:

$$p(\hat{\psi}_j^2(t_2)\,\&\,\hat{\psi}_i^1(t_2)) \approx |\mu_i|^2\,|\nu_j^i|^2 \tag{18}$$

(the equality is only approximate, because the possessed properties are not exactly the $|\psi_j^2\rangle$ and the $|\psi_j^1\rangle$. Finally,

$$p(\hat{\psi}_j^2(t_2)|\hat{\psi}_i^1(t_1)) \approx \frac{|\mu_i|^2|\nu_j^i|^2}{|\mu_i|^2} = |\nu_j^i|^2, \tag{19}$$

which is equal to $|\langle\phi_i^*|\eta_j\rangle|^2$, that is, the Born probability. Notice that by virtue of the Vermaas-Dieks correlation rule, the calculated probability depends only on the reduced state of the system $M_1 \otimes M_2$, and in particular does *not* depend on the possessed properties of the measured system.

If we allow time to pass between the preparation and the registration, the situation is only slightly more involved. Since the system no longer interacts with M_1, the evolution of the state factorizes as the product of the free evolution of the system and the evolution of the rest (which is a decoherence evolution that leaves the states of M_1 and M_2 invariant). So, the only significant difference is that each of the vectors $|\phi_i^*\rangle$ has approximately evolved according to the Schrödinger equation for the system S:

$$|\phi_i^*\rangle \mapsto e^{-iHt}|\phi_i^*\rangle, \tag{20}$$

and thus the conditional probabilities according to the modal interpretation will be approximately given by $|\langle\phi_i^*|e^{iHt}|\eta_j\rangle|^2$. These are simply the Born probabilities calculated from the Schrödinger evolved state of the system according to the standard theory.

It is now clear that we can also relax the assumption that the second measurement interaction represents a measurement of a self-adjoint operator and has the form (11). The treatment of the general case is straightforward and similar to the above. In fact, the assumption of *definiteness* of pointer readings guarantees that the definite properties of the preparation and registration devices always represent definite pointer readings of the devices. Further, the Vermaas-Dieks correlation rule implies that in any situation the probabilities for the joint possession of properties will be the same as the Born probabilities for the results of joint measurements of the observables corresponding to the possessed properties. Thus, the modal interpretation reproduces the correlations predicted in standard quantum mechanics between the readings of preparation and registration devices at any one time quite irrespective of the details of the interactions. The further assumption of *stability* allows us then to derive the corresponding *two-time* transition probabilities from these *single-time* joints. Notice that we do not need any details of the dynamics to derive our results.

One may disagree as to whether stability in this sense is a necessary requirement. One can argue that even in the standard theory, all that matters are the correlations between the outcomes of the registration at t_2 and the records at t_2 of the outcomes of the preparation. According to this view, which has been famously discussed by Bell (1976) in the context of the many-worlds interpretation, the predictions of standard quantum mechanics are recovered as soon as the predicted instantaneous correlations at t_2 are the same in the modal interpretation and in the standard theory. In this sense, it would be immaterial whether the records at t_2 are faithful records of the outcomes $|\psi_i^1(t_1)\rangle$ of the preparation. We find, however, the assumption of stability to be more satisfactory. Further, stability for the models used here is indeed satisfied in the main proposal for dynamics given in Bacciagaluppi and Dickson (1997).

Thus, we have satisfied both *desiderata* (A) and (B). The properties of the preparation device M_1 play a crucial role in both cases. The subensembles selected in measurements are characterized not by the properties of the system but by the properties of the preparation device M_1, which thus define the "collapsed state" of the system. Similarly, the Born rule is recovered by using the probabilistic dependence of the outcomes of the registration on the possessed properties of M_1. In other words, the possessed properties of M_1 explain the role of the collapsed state $|\phi_j^*\rangle$ as an *effective* state. (See also the related discussion in Vermaas [1996].)

4. The Modal Interpretation as a Stochastic Hidden-Variables Theory

What is usually meant by a hidden-variables theory is the following: the hidden variables of the object system (together with its quantum state and, in general, a measurement context) determine, maybe only with a certain probability, the outcomes of a measurement, in such a way that if one averages over the hidden variables, one recovers the quantum-mechanical statistics given by the quantum state of the system, where "state" here and above means the collapsed state of standard quantum theory.

In the modal interpretation, possessed properties have to evolve according to some stochastic dynamics, and important aspects of the dynamics can be encoded in two-time transition probabilities, like the ones studied by Vermaas (1996). This allows one to investigate the *probabilistic dependence* of the properties of some system on previously possessed properties of the same or other systems. In particular, this means that we can study the probabilistic dependence of measurement results, understood as possessed properties of a measuring apparatus, on the possessed properties of the measured system, which we can understand as *hidden variables*.

We have seen that outcomes of measurements depend probabilistically on previously possessed properties of preparation devices, and that the probabilistic predictions of quantum mechanics for outcomes of measurements can be recovered on the basis of this probabilistic dependence, without reference to the properties of the measured system. This does *not* mean, however, that the outcomes of measurements are probabilistically independent of the possessed properties of the system. On the contrary, one naturally expects that the outcomes $|\hat{\psi}_j(t_2)\rangle$ of a measurement will depend also on the possessed properties $|\hat{\phi}_i(t_1)\rangle$ of the system. This is the case in the following example in which, without having to solve the full dynamical equations, one can show genuine probabilistic dependence of the measurement results both on the possessed properties of the system and on the effective state (although the dependence on the effective state turns out to be screened off).

Take the system to be a composite $S = S_1 \otimes S_2$, and let the state $|\Psi_1\rangle$ at the time t_1 when the preparation is completed be

$$\sum_i \mu_i |\phi_i^*\rangle \otimes |\psi_i^1\rangle \otimes |E_{i0}\rangle. \tag{21}$$

In general, the state $|\phi^*_i(t_1)\rangle$ of S is an entangled state, and the corresponding state of system S_2 (or S_1) is an improper mixture. We now assume that the possessed properties of the system S have the form:

$$|\hat{\phi}_i(t_1)\rangle = |\alpha_i\rangle \otimes |\beta_i\rangle, \tag{22}$$

that is, they are *product properties* (which we assume to be one-dimensional). The properties of S_2 will thus be the states $|\beta_i\rangle$, and they will be perfectly correlated to the properties $|\alpha_i\rangle$ of S_1.

Now let the registration be a measurement of an arbitrary observable *of the system* S_2, with eigenvectors

$$|\eta_i\rangle = \sum_j \nu^i_j |\beta_j\rangle. \tag{23}$$

During the registration, we assume that S_1 does not interact with other systems. Under this assumption, Vermaas (1996) has shown that the possessed properties of S_1 will evolve deterministically according to the free Heisenberg evolution of S_1 (the same is true for the dynamics of Bacciagaluppi and Dickson [1997]). Thus, conditionalizing on the possessed properties of S_1 at t_1 is equivalent to conditionalizing on the possessed properties of S_1 at t_2.

The final state at t_2 when the registration is complete is:

$$|\Psi_2\rangle = \sum_{i,j,k} \mu_i \lambda^i_j \nu^j_k |\alpha_j\rangle \otimes |\eta_k\rangle \otimes |\psi^1_i\rangle \otimes |\psi^2_k\rangle \otimes |E_{ik}\rangle. \tag{24}$$

From this state, using the perfect correlations between the properties of S_1 and S_2 at t_1 and the deterministic evolution (or stability, respectively) of S_1 and M_1, as well as the fact that

$$\sum_j |\lambda^i_j|^2 = \sum_k |\nu^j_k|^2 = 1, \tag{25}$$

we can calculate the following probabilities:

$$\begin{aligned}
\mathrm{p}(\hat{\psi}^2_k(t_2)|\hat{\psi}^1_i(t_1)) &= \frac{\Sigma_j\, |\mu_i|^2|\lambda^i_j|^2|\nu^j_k|^2}{\Sigma_{j,k}|\mu_i|^2|\lambda^i_j|^2|\nu^j_k|^2} = \sum_j |\lambda^i_j|^2|\nu^j_k|^2, \\
\mathrm{p}(\hat{\psi}^2_k(t_2)|\hat{\psi}^1_i(t_1)\&\beta_j(t_1)) &= \frac{|\mu_i|^2|\lambda^i_j|^2|\nu^j_k|^2}{\Sigma_k|\mu_i|^2|\lambda^i_j|^2|\nu^j_k|^2} = |\nu^j_k|^2, \\
\mathrm{p}(\hat{\psi}^2_k(t_2)|\beta_j(t_1)) &= \frac{\Sigma_i|\mu_i|^2|\lambda^i_j|^2|\nu^j_k|^2}{\Sigma_{i,k}|\mu_i|^2|\lambda^i_j|^2|\nu^j_k|^2} = |\nu^j_k|^2.
\end{aligned} \tag{26}$$

The last of these probabilities, the one conditional on the possessed properties of the system, has the form

$$\mathrm{p}(\hat{\psi}^2_k(t_2)|\beta_j(t_1)) = |\nu^j_k|^2 = |\langle\eta_k|\beta_j\rangle|^2, \tag{27}$$

by (23), that is, it has the same form as the Born rule. This conforms to a result by Vermaas (1996) and depends on the fact that there is a system, namely S_1, whose properties *before* the measurement are perfectly correlated with the properties of the system to be measured. In general, as Vermaas (1996) explicitly shows, transition probabilities will not be Born-like.

The example illustrates how in general conditionalizing upon both the possessed property of S_2 and the effective state yields more information than conditionalizing only on the effective state; in this example, further, conditionalizing on the possessed property of S_2 screens off dependence on the effective state, even if it does not determine the result of the measurement with probability 1.

Thus, in general, the measurement outcomes depend probabilistically not only on the effective state of the system, which corresponds to a possessed property of the preparation device, but also on a property of the system that is not directly accessible. That is, the measurement outcome depends (stochastically) both on the effective state of the system, and on some additional *hidden variables* of the system. If we average over the hidden variables, only the dependence on the effective state remains.

5. Holism and Nonlocality

In this final section, we shall attempt to connect the discussion of the possessed properties of the system with the issue of nonlocality of hidden variables (see also Dickson and Clifton, [1998]). In particular, we shall link to Healey's (1989, 1994) discussion of holism and nonlocality in the modal interpretation.

In standard quantum mechanics, the (collapsed) states ascribed to systems play an important explanatory role, because in a collapse interpretation (as we understand it), properties are possessed if and only if the corresponding projections have expectation value 0 or 1. Once the state of a quantum system has collapsed, the system has acquired certain properties. These then evolve according to the Schrödinger equation of the system, if it does not interact with other systems in the meantime, and then these properties explain why, upon measurement, certain outcomes will be obtained with certain probabilities.

Consider in particular the usual EPR (or EPR-Bohm) situation. Take two spatially separated electrons, say S_1 and S_2, in the singlet state $|\Psi_0\rangle$, which has the form

$$|\Psi_0\rangle = \frac{1}{\sqrt{2}} (|+_r\rangle \otimes |-_r\rangle + |-_r\rangle \otimes |+_r\rangle), \tag{28}$$

with respect to any of the bases $(|+_r\rangle, |-_r\rangle)$ in the Hilbert spaces of S_1 and S_2, where $|+_r\rangle$ and $|-_r\rangle$ are the eigenstates of the component of spin in the direction r. In order to avoid exact degeneracy, we can consider the electron pair to be in a state $|\Psi\rangle$ that is close to the singlet state, say $\|\Psi - \Psi_0\|^2 < \epsilon^2$. In this case, with respect to the bases $(|+_r\rangle, |-_r\rangle)$, $|\Psi\rangle$ takes the form

$$|\Psi\rangle = \lambda^r_{+-}|+_r\rangle \otimes |-_r\rangle + \lambda^r_{-+}|-_r\rangle \otimes |+_r\rangle + \\ + \lambda^r_{++}|+_r\rangle \otimes |+_r\rangle + \lambda^r_{--}|-_r\rangle \otimes |-_r\rangle, \quad (29)$$

where now the coefficients depend in general on the direction r, but for all r,

$$|\tfrac{1}{\sqrt{2}} - \lambda^r_{+-}|^2 + |\tfrac{1}{\sqrt{2}} - \lambda^r_{-+}|^2 + |\lambda^r_{++}|^2 + |\lambda^r_{--}|^2 = \\ \text{i}= \|\Psi - \Psi_0\|^2 < \epsilon^2. \quad (30)$$

That is, the (anti)correlation between the spin components of S_1 and S_2 in any direction is still very high.

Since the actual measurement interactions for measurements on S_1 and S_2, respectively, are obviously local (the magnetic fields in a Stern-Gerlach measurement need to act only on the *single* electrons), these nonlocal correlations are famously what needs to be explained. In the standard theory, they are explained by the nonlocal nature of the collapse, by which, say, upon an ideal measurement with outcome $|-_r\rangle$ of the spin of S_1 in the direction r, the state of the electron *pair* collapses to

$$\lambda^r_{-+}|-_r\rangle \otimes |+_r\rangle + \lambda^r_{--}|-_r\rangle \otimes |-_r\rangle \quad (31)$$

(up to a normalization factor), where the second term is of order ϵ in norm.

In the modal interpretation the collapse does not take place, and, further, by standard nonsignalling results, the reduced state of S_2 remains invariant during a measurement on S_1. Indeed, an interaction that is localized at S_1 cannot change the reduced state of S_2 (see, for example, Redhead [1987, 114, 118]). Thus, in the modal interpretation, the set of definite properties of S_2 remains *unchanged* by the act of measurement on S_1.[5]

On the one hand, there are directions r_1 and r_2 for which the above state $|\Psi\rangle$ takes the form

$$|\Psi\rangle = \lambda_1|+_{r_1}\rangle \otimes |-_{r_2}\rangle + \lambda_2|-_{r_1}\rangle \otimes |+_{r_2}\rangle. \quad (32)$$

This is simply by the biorthogonal decomposition theorem, together with the fact that for a spin-$\frac{1}{2}$ particle every basis in the Hilbert space has the form $(|+_r\rangle, |-_r\rangle)$ for some r. Further, the directions r_1 and r_2 are almost parallel (because $|\Psi\rangle$ is close to the singlet state). The perfect (anti)correlations obtained in the measurements of spin in the directions

r_1 and r_2 on S_1 and S_2, respectively, are no mystery in the modal interpretation, because these spin components are already possessed properties, and it is always assumed (and can be shown in the full dynamics; see Bacciagaluppi [1996a]) that ideal measurements of possessed properties are faithful.

On the other hand, if some other measurements are performed on the two wings, say both in a direction r significantly different from r_1 and r_2, then we have seen that under the assumptions of definiteness and stability of pointer readings, the modal interpretation will reproduce the quantum mechanical statistics, including the almost perfect correlations. Now, indeed, these correlations may at first appear quite puzzling, because we seem to have a direct probabilistic dependence of the outcomes of the second measurement (on S_2) on the outcomes of the first (on S_1), that essentially screens off the dependence on the actual hidden variables of S_2.

One way of explaining the nonlocal correlations was suggested by Healey (1989, chapter 4 and esp. chapter 5; 1994, sections VII–VIII), and is based on the ascription of *holistic properties* to the system $S_1 \otimes S_2$.[6] Healey's suggestion is easy to understand in light of the above discussion of state preparation. In fact, we can consider the first measurement, performed by an apparatus M_1, as a measurement not simply on S_1, but on the electron pair $S_1 \otimes S_2$, and we can consider it a *preparation procedure*. The total state of the system $M_1 \otimes S_1 \otimes S_2$ after this first measurement will have the form

$$\mu_+ |\phi_+^*\rangle \otimes |\psi_+^1\rangle + \mu_- |\phi_-^*\rangle \otimes |\psi_-^1\rangle. \tag{33}$$

Now suppose, as assumed by Healey in his discussion, that

$$\langle \phi_+^* | \phi_-^* \rangle = 0. \tag{34}$$

For instance, the M_1-measurement could be an ideal measurement of the spin of S_1 in the direction r, in which case the states $|\phi_\pm^*\rangle$ are simply the two states

$$\lambda_{+\pm}^r |+_r\rangle \otimes |\pm_r\rangle + \lambda_{-\pm}^r |-_r\rangle \otimes |\pm_r\rangle \tag{35}$$

(up to normalization), which are indeed orthogonal.

The appeal of this suggestion is that, since the $|\phi_\pm^*\rangle$ are orthogonal, if we assign a possessed property to the *composite* system $S_1 \otimes S_2$, it will be the projection onto the collapsed state $|\phi_\pm^*\rangle$. Thus, one can think of the correlations between the outcomes of the M_2-measurement and the M_1-measurement as mediated by the holistic properties (the hidden variables) of the composite, as in the analogous case in the standard formulation of quantum mechanics, in which they are mediated by the collapsed state $|\phi_\pm^*\rangle$. This explanation is enhanced by the fact that, as in our previous analysis, the outcomes of the M_2-measurement exhibit the "correct" probabilistic dependence on the holistic properties of $S_1 \otimes S_2$. Healey's proposed explanation of the correlations can be considered as a local process, involving as it does only the local interaction between the system S and the apparatus M_2 (although S is a spacelike extended system). Healey's motivation for stressing the role of the holistic properties lies precisely in the wish to dispense with nonlocal processes.[7]

Yet this is not the most general case, and our analysis of state preparation also indicates the limits of this approach. In fact, if the M_1-measurement is such that the vectors $|\phi^*_\pm\rangle$ in (33) are *not* orthogonal (and this can, indeed, be reached through a local measurement interaction), then we have shown in sections 2 and 3 that the probabilistic dependence on the properties of $S_1 \otimes S_2$ will not be enough to recover the quantum-mechanical statistics, and that the probabilistic dependence on the properties of M_1 is essential. One could reply that this dependence could be, again, explained in terms of holistic properties, namely, the holistic properties of the composite $S_1 \otimes S_2 \otimes M_1$.[8] It seems to us, however, that there is a substantial difference between invoking holistic properties of a microscopic, if delocalized, quantum system such as the electron pair, in order to explain the quantum-mechanical correlations, and invoking holistic properties of a composite of a microscopic system and a macroscopic piece of apparatus. We think that committing oneself to this extreme form of holism would be a rather high price to pay.

The alternative way of explaining the correlations is to fill out the details of the *dynamical process* that leads to them. This is precisely analogous to the explanatory force of the de Broglie-Bohm theory, which postulates that positions of all systems are distributed according to the quantum-mechanical distribution, but also shows how this distribution is maintained by the dynamical evolution of the theory (is equivariant). Thus it provides a detailed mechanism by which, in particular, positions of pointers come to be distributed as they are, including their dependence on positions of other pointers.

Dickson and Clifton (1998) have shown that any dynamics that satisfies stability, irrespective of any details, will be nonlocal in the strong sense of violating fundamental Lorentz invariance. They show that, given the possessed properties at the source, the probabilities for (and thus the proportion of) different outcomes of the measurements on the two wings of the experiment will be *frame dependent*. By the above, it is already clear that the modal interpretation violates the Bell inequalities, and our discussion of Healey's proposal shows that it exhibits *outcome dependence*. It is not difficult to reformulate Dickson and Clifton's (1998) results using the assumption of parameter independence instead of fundamental Lorentz invariance, and thus show that the modal interpretation exhibits also *parameter dependence*. Thus the alternative is between full-blown holism and full-blown nonlocality.

As a final remark, we mention that there are independent reasons for wanting to relinquish the idea of holistic properties, represented by the no-go theorems of Bacciagaluppi (1995), Clifton (1996), and Vermaas (1997). The resulting so-called "atomic" version of the modal interpretation, in which properties are assigned only to a layer of "atomic" systems and to none of their composites, is considered in this same volume in the chapter by Dennis Dieks.[9]

Notes

We heartily thank Richard Healey and Geoffrey Hellman, together with the staff at the Minnesota Center for the Philosophy of Science, for organizing a most productive and enjoyable meeting and for securing financial support. We are indebted first and foremost to Jeremy Butterfield, and also to Michael Dickson and Pieter Vermaas. Further, we are grateful to David Albert, Dennis Dieks, Michael Redhead, Laura Ruetsche, and Simon Saunders, as well as to the audience at the University of Minnesota. G. B. acknowledges support from the British Academy and the Arnold Gerstenberg Fund, and M. H. from the Cambridge Overseas Trust and the British Overseas Research Scheme.

1. Bub's proposed framework is that of *beable theories* for quantum mechanics. An important aspect of such theories is their strong analogy with the de Broglie-Bohm theory (Bub, 1996, 1997). The sense in which (some) modal interpretations can also be thought of as beable theories and the way to develop a corresponding de Broglie-Bohm dynamics is detailed in Bacciagaluppi and Dickson (1997).

2. More precisely, it is the state of the composite consisting of system and apparatus that collapses. The basis for the collapse is always determined by the eigenstates of the pointer observable, both in the case of ideal and nonideal measurements. If the pointer observable is nonmaximal (coarse-grained), it has multidimensional eigenspaces, and the state of the system will be a mixture of states that are clearly themselves improper mixtures. This last situation appears also in the example at the end of section 4.

3. Not all measurement interactions correspond to measurements of self-adjoint operators. In Bacciagaluppi and Hemmo (1996) we treat the general case, and the results derived here can be extended straightforwardly. The theory that deals with general measurements is known as unsharp quantum theory; for a textbook exposition, see Busch, Grabowski, and Lahti (1995).

4. We are assuming for simplicity that the pointer states $|\psi_i^1\rangle$ remain stationary under the total time evolution after the preparation is completed. This implies no loss of generality: if the pointer states are not stationary (if the apparatus is moved to another location or the pointer gets rusty, for example), then the $|\hat{\psi}_i^1(t_2)\rangle$ will be close to the time evolutes from t_1 to t_2 of the $|\psi_i^1\rangle$. Similarly, M_1 could interact with some other system, causing the pointer reading to be recorded in some other form (e.g., a computer memory). In this case, one would redefine the $|\hat{\psi}_i^1(t_2)\rangle$ accordingly.

5. A different question is whether the *actually possessed* property of S_2 remains unchanged. Again, in the case of the dynamics by Bacciagaluppi and Dickson (1997), the answer is that it remains indeed unchanged.

6. Notice that the properties assigned to a composite system according to the spectral decomposition of its reduced state will generally not be product properties; in particular, they will not be the products of the properties of its component systems, and they will not commute with the properties of the components. Thus, the properties of the parts do not determine the properties of the whole, and in general the properties of a composite are, indeed, holistic properties.

7. It should be stressed that, as a matter of fact, probabilistic dependence is exhibited on *both* the properties of S and on the properties of M_1. (In fact, the same dependence is exhibited also on the properties of $S \otimes M_1$.) Probabilistic dependence by itself does not carry a commitment to causal influence, however. One could presumably maintain that only the probabilistic dependence on, say, properties of S indicates a direct causal link, but not the dependence on properties of M_1.

8. In fact, since both the properties of M_1 and the properties of $S_1 \otimes S_2 \otimes M_1$ are (almost) perfectly correlated with the properties of the environment, by the Vermaas-Dieks correlation rule, it follows that the properties of M_1 and of $S_1 \otimes S_2 \otimes M_1$ are almost perfectly correlated. Further, the holistic properties of $S_1 \otimes S_2 \otimes M_1$, although they are not in general product properties (and in particular not the products of the properties of M_1 and $S_1 \otimes S_2$), will be *close* to the product states $|\phi_i^*\rangle \otimes |\psi_i^1\rangle$. Thus, in a spirit close to Healey's, one could use these instead of the properties of M_1, both for the selection of the appropriate subensembles and for the recovery of the appropriate probabilistic dependence.

9. See also Bacciagaluppi and Dickson (1997), Vermaas (1998), and Bacciagaluppi and Vermaas (1998). The latter paper also contains a brief review of the no-go results.

References

Albert, D., and B. Loewer. 1990. "Wanted Dead or Alive: Two Attempts to Solve Schrödinger's Paradox." In A. Fine, M. Forbes, and L. Wessels, eds., *Proc. 1990 Biennial Meeting of the Philosophy of Science Association*, vol. 1. East Lansing: Philosophy of Science Association, 277–85.

Bacciagaluppi, G. 1995. "Kochen-Specker Theorem in the Modal Interpretation of Quantum Mechanics." *International Journal of Theoretical Physics* 34: 1206–15.

———. 1996a. "Topics in the Modal Interpretation of Quantum Mechanics." Doctoral dissertation, Cambridge University.

———. 1996b. "Delocalised Properties in the Modal Interpretation of a Continuous Model of Decoherence." Preprint.

Bacciagaluppi, G., and M. Dickson. 1996. "Dynamics for Density Operator Interpretations of Quantum Theory," quant-ph/9711048.

Bacciagaluppi, G., and M. Hemmo. 1996. "Modal Interpretations, Decoherence, and Measurements." *Studies in the History and Philosophy of Modern Physics* 27: 239–77.

Bacciagaluppi, G., and P. E. Vermaas. 1998. "Virtual Reality: Consequences of No-Go Theorems for the Modal Interpretation of Quantum Mechanics." Forthcoming in M. L. Dalla Chiara, R. Giuntini, and F. Laudisa, eds., *Philosophy of Science in Florence, 1995*. Dordrecht: Kluwer.

Bacciagaluppi, G., M. J. Donald, and P. E. Vermaas. 1995. "Continuity and Discontinuity of Definite Properties in the Modal Interpretation." *Helvetica Physica Acta* 68: 679–704.

Bacciagaluppi, G., A. Elby, and M. Hemmo. 1996. "Measurement Problem and Observers' Beliefs in the Modal Interpretation of Quantum Mechanics." Preprint.

Bell, J. S. 1976. "The Measurement Theory of Everett and de Broglie's Pilot Wave." In J. S. Bell, *Speakable and Unspeakable in Quantum Mechanics*. Cambridge: Cambridge University Press, 1987, 93–99.

Bub, J. 1992. "Quantum Mechanics without the Projection Postulate." *Foundations of Physics* 22: 737–54.
———. 1996. "Schrödinger's Cat and Other Entanglements of Quantum Mechanics." Forthcoming in J. Earman and J. Norton, eds., *The Cosmos of Science*. Pittsburgh: University of Pittsburgh Press; Konstanz: Universitäts-Verlag Konstanz.
———. 1997. *Interpreting the Quantum World*. Cambridge: Cambridge University Press.
Busch, P., M. Grabowski, and P. Lahti. 1995. *Operational Quantum Physics*. Lecture Notes in Physics. Berlin: Springer.
Clifton, R. 1995. "Independently Motivating the Kochen-Dieks Modal Interpretation of Quantum Mechanics." *British Journal for the Philosophy of Science* 46: 33–57.
———. 1996. "The Properties of Modal Interpretations of Quantum Mechanics." *British Journal for the Philosophy of Science* 47: 371–98.
Dickson, M., and R. Clifton. 1998. "Lorentz-Invariance in the Modal Interpretation." Forthcoming in Dieks and Vermaas (1998).
Dieks, D. 1989. "Resolution of the Measurement Problem Through Decoherence of the Quantum State." *Physics Letters* A142: 439–46.
———. 1994. "The Modal Interpretation of Quantum Mechanics, Measurement, and Macroscopic Behaviour." *Physical Review* D49: 2290–2300.
———. 1995. "Physical Motivation of the Modal Interpretation of Quantum Mechanics." *Physics Letters* A197: 367–71.
Dieks, D., and P. E. Vermaas, eds. 1998. *The Modal Interpretation of Quantum Mechanics*. Dordrecht: Kluwer, forthcoming.
Donald, M. J. 1998. "Discontinuity and Continuity of Definite Properties in the Modal Interpretation." Forthcoming in Dieks and Vermaas (1998).
Healey, R. 1989. *The Philosophy of Quantum Mechanics: An Interactive Interpretation*. Cambridge: Cambridge University Press.
———. 1994. "Nonseparable Processes and Causal Explanation." *Studies in the History and Philosophy of Modern Physics* 25: 337–74.
Kochen, S. 1985. "A New Interpretation of Quantum Mechanics." In P. Lahti and P. Mittelstaedt, eds., *Symposium on the Foundations of Modern Physics*. Singapore: World Scientific, 151–69.
Krips, H. 1987. *The Metaphysics of Quantum Theory*. Oxford: Clarendon Press.
Redhead, M. L. G. 1987. *Incompleteness, Nonlocality, and Realism: A Prolegomenon to the Philosophy of Quantum Mechanics*. Oxford: Clarendon Press.
Ruetsche, L. 1995. "Measurement Error and the Albert-Loewer Problem." *Foundations of Physics Letters* 8: 327–44.
van Fraassen, B. C. 1973. "Semantic Analysis of Quantum Logic." In C. A. Hooker, ed., *Contemporary Research in the Foundations and Philosophy of Quantum Theory*. Dordrecht and Boston: Reidel, 180–213.
———. 1991. *Quantum Mechanics: An Empiricist View*. Oxford: Clarendon Press.
Vermaas, P. E. 1996. "Unique Transition Probabilities in the Modal Interpretation." *Studies in the History and Philosophy of Modern Physics* 27: 133–59.
———. 1997. "A No-Go Theorem for Joint Property Ascriptions in the Modal Interpretation of Quantum Mechanics." *Physical Review Letters* 78: 2033–37.
———. 1998. "The Pros and Cons of the Kochen-Dieks and the Atomic Modal Interpretation." Forthcoming in Dieks and Vermaas (1998).
Vermaas, P. E., and D. Dieks. 1995. "The Modal Interpretation of Quantum Mechanics and Its Generalization to Density Operators." *Foundations of Physics* 25: 145–58.
Zurek, W. H. 1993. "Preferred States, Predictability, Classicality, and the Environment-Induced Decoherence." *Progress of Theoretical Physics* 89: 281–312.

Pieter E. Vermaas

Expanding the Property Ascriptions in the Modal Interpretation of Quantum Theory

1. Introduction

Without a sound policy it is tricky business to assign definite values to quantum-mechanical magnitudes. As Kochen and Specker (1967) have shown, a value assignment to all magnitudes of a quantum system can lead to contradictions. The modal interpretation of quantum theory assigns definite values to a limited set of magnitudes without contradictions. It is my aim to expand this limited set as much as possible.

Take a system α with a state represented by a density operator W, and let the projections $\{P_j\}_j$ be the eigenprojections of W. Then, in the modal interpretation in the version[1] by Vermaas and Dieks (1995), there is probability $\mathrm{Tr}(P_k W)$ that the magnitude represented by P_k has value 1 and that the magnitudes represented by the other eigenprojections $\{P_j\}_{j \neq k}$ have all value 0.

This value assignment to the magnitudes represented by $\{P_j\}_j$ is a *core* value assignment and induces an assignment of values to other magnitudes. Consider magnitudes represented by operators $A = \Sigma_j\, a_j P_j$, in other words, magnitudes represented by operators that have the same eigenprojections as the state W. Then, if the magnitude represented by P_k has value 1, a magnitude represented by such an operator A has as value the eigenvalue a_k corresponding to P_k.

The question I address in this chapter is whether one can assign definite values to more magnitudes than just those represented by the operators $A = \Sigma_j\, a_j P_j$. My aim is to develop a sound policy for finding the largest possible set of definite-valued magnitudes without running into contradictions.

I proceed in two steps. First, I determine the largest possible set of definite-valued magnitudes represented by projections. Value assignments to projections have been discussed before, especially by Kochen (1985) and Clifton (1995a). The full treatment of value assignments to projections will take up the larger part of this chapter. Second, I discuss how the value assignment to magnitudes represented by projections induces a value assignment to magnitudes represented by arbitrary operators.

An assignment of values 0 or 1 to projections $\{Q\}$ can equivalently be taken as a truth-value assignment to propositions about possessed properties. The assignment of value 1 to Q implies that the proposition "α possesses the property represented by Q" is true; the assignment of 0 to Q implies that this proposition is false. Value assignments to projections can thus be regarded as property ascriptions and the core value assignment to the eigenprojections can thus be regarded as the *core property ascription*. In section 2 I start by presenting the core property ascription.

The core property ascription can be extended to what I call the *expanded property ascription*. Let $\mathcal{DP}(W)$ be the set of projections that, given a state W, have definite values according to the expanded property ascription ($\mathcal{DP}$ stands for definite-valued projections).

Kochen was the first to propose a determination of $\mathcal{DP}$. Clifton later put forward an alternative proposal. In sections 3 and 4 Kochen's and Clifton's expanded property ascriptions, denoted by $\mathcal{DP}_{\mathrm{K}}$ and $\mathcal{DP}_{\mathrm{C}}$, respectively, are discussed. Clifton assigns definite values to more projections than Kochen does, so, given the aim of this paper, $\mathcal{DP}_{\mathrm{C}}$ is preferred over $\mathcal{DP}_{\mathrm{K}}$. But maybe other property ascriptions exist that assign definite values to even more projections. There are arguments against and in favor of the existence of such other property ascriptions.

Clifton (1995a; 1995b) and Dickson (1995) proposed conditions that a satisfactory property ascription should obey. The only property ascription that meets these conditions is $\mathcal{DP}_{\mathrm{C}}$. As a consequence, $\mathcal{DP}_{\mathrm{C}}$ seems to be the only satisfactory property ascription. In section 5 one of Clifton's arguments is briefly discussed.

Healey (1989) and Arntzenius (1990) proposed other conditions that a property ascription should obey; see sections 6 and 7. The sets $\mathcal{DP}_{\mathrm{K}}$ and $\mathcal{DP}_{\mathrm{C}}$ do not meet these conditions. As a consequence, $\mathcal{DP}_{\mathrm{C}}$ appears not to be a satisfactory property ascription. Further, if one puts all the above conditions together and demands that the expanded property ascription meet them all, one is faced with a contradiction: If $\mathcal{DP}$ meets Clifton's and Dickson's conditions, $\mathcal{DP}$ is $\mathcal{DP}_{\mathrm{C}}$ and $\mathcal{DP}$ cannot meet Healey's and Arntzenius's conditions; if $\mathcal{DP}$ meets Healey's and Arntzenius's conditions, $\mathcal{DP}$ is not $\mathcal{DP}_{\mathrm{C}}$ and $\mathcal{DP}$ cannot meet Clifton's and Dickson's conditions.

In section 8 I develop a policy to determine a new expanded property ascription. This new expanded property ascription defines two different sets of definite-valued projections. The first is a set of projections that have possibly—with a probability >0—definite values. The second is a set of projections that have always—with probability 1—definite values. I call these sets $\mathscr{DP}_{\text{Prob}>0}(W)$ and $\mathscr{DP}_{\text{Prob}=1}(W)$, respectively.

The new expanded property ascription resolves the above contradiction. This contradiction arose because all the proposed conditions were imposed on one and the same set of definite-valued projections. But because the new expanded property ascription yields two sets of definite-valued projections, this is no longer necessary. In section 9 I show that Clifton's and Dickson's conditions should only be imposed on $\mathscr{DP}_{\text{Prob}=1}$ and that Healey's and Arntzenius's conditions should only be imposed on $\mathscr{DP}_{\text{Prob}>0}$. The new expanded property ascription meets in this way all proposed conditions. Furthermore, the new property ascription is larger than Clifton's property ascription in the sense that $\mathscr{DP}_{\text{Prob}>0}$ is larger than $\mathscr{DP}_{\text{C}}$.

Finally, in section 10, I discuss which magnitudes not represented by projections have definite values.

2. The Core Property Ascription

Take a quantum system α with associated Hilbert space $\mathscr{H}^{\alpha}$ and suppose that the state of α is given at time t by a density operator W defined in $\mathscr{H}^{\alpha}$.

The modal interpretation in the version by Vermaas and Dieks (1995) assigns values to the projections which occur in the spectral resolution of the state W. Let $[Q]$ be the value of a projection Q. Let the spectral resolution of W be[2]

$$W = \sum_j w_j P_j.$$

Then, according to the modal interpretation, the eigenprojections $\{P_j\}_j$ have definite values. These values are determined probabilistically: with probability $\text{Tr}(P_k W)$ the values of the eigenprojections are simultaneously equal to

$$[P_k] = 1 \quad \text{and} \quad [P_j] = 0 \quad \text{if} \quad j \neq k.$$

Given this core property ascription, one can look for the expanded property ascription.

3. Kochen's Expanded Property Ascription

Kochen (1985) was the first to propose a way to determine $\mathscr{DP}$.[3] According to Kochen $\mathscr{DP}(W)$ is the Boolean σ-algebra generated by the eigenprojections $\{P_j\}_j$ of the state W. Let $\mathcal{B}(\{P_j\}_j)$ signify this Boolean algebra. Kochen's property ascription is thus[4]

$$\mathcal{DP}_K(W) := \mathcal{B}(\{P_j\}_j).$$

Kochen motivates his determination of $\mathcal{DP}$ by emphasizing that in this way the property ascription inherits a classical structure: if Q has a definite value, then so does its negation $\neg Q$, and if Q and $\tilde{Q}$ have definite values, then so do their conjunction and disjunction.

Kochen's proposal can be understood as follows. Start with the core property ascription and demand that $\mathcal{DP}(W)$ is closed under the lattice operations join $\vee$, meet $\wedge$, and orthocomplementation $\neg$.[5] Call this condition

Closure:
If $Q, \tilde{Q} \in \mathcal{DP}(W)$, then also $Q \vee \tilde{Q}$, $Q \wedge \tilde{Q}$, $\neg Q \in \mathcal{DP}(W)$.

The closure of the set $\{P_j\}_j$ is the Boolean algebra $\mathcal{B}(\{P_j\}_j)$. Kochen's proposal can thus be summarized as

$$\left.\begin{array}{c}\text{Core Property Ascription}\\ +\ \text{Closure}\end{array}\right\} \rightarrow \mathcal{DP}(W) = \mathcal{B}(\{P_j\}_j).$$

4. Clifton's Expanded Property Ascription

Clifton (1995a) also proposed a way to determine $\mathcal{DP}$.[6] Clifton gives two equivalent definitions. Let $\mathcal{N}(W)$ be the set of projections onto subspaces of the null-eigenspace of W. The first definition is

$$\mathcal{DP}_C(W) = \{Q \mid Q = Q_1 + Q_2,\ Q_1 \in \mathcal{B}(\{P_j\}_j),\ Q_2 \in \mathcal{N}(W)\}. \quad (4.1)$$

It follows that $\mathcal{DP}_C$ is larger than Kochen's property ascription: all members of $\mathcal{DP}_K(W)$ are also members of $\mathcal{DP}_C(W)$, but if $\mathcal{N}(W)$ is not the empty set, the converse does not hold. The second definition is

$$\mathcal{DP}_C(W) = \{Q \mid QP_j = P_j \text{ or } 0, \text{ for any } P_j \text{ in } \{P_j\}_j\}. \quad (4.2)$$

For the proof that the definitions (4.1) and (4.2) are equivalent, see Clifton (1995a).

Dickson (1995) characterized $\mathcal{DP}_C$ as a *faux-Boolean algebra*. Take a set of mutual orthogonal projections $\{S_j\}_j$. This set generates a set of projections that are orthogonal to all S_j, that is, the set $\{Q \mid Q \sum_j S_j = 0\}$. The faux-Boolean algebra generated by $\{S_j\}_j$ is now the closure under the join $\vee$, meet $\wedge$, and orthocomplementation $\neg$, of the union of $\{S_j\}_j$ and $\{Q \mid Q \sum_j S_j = 0\}$.[7] Let $\mathcal{F}(\{S_j\}_j)$ signify this faux-Boolean algebra.

It is easily seen that $\mathcal{DP}_C(W)$ is the faux-Boolean algebra generated by the eigenprojections of W, so

$$\begin{aligned} \mathcal{DP}_C(W) &= \mathcal{F}(\{P_j\}_j) \\ &= \text{Closure}_{\vee,\wedge,\neg} \text{ of } \{P_j\}_j \cup \left\{Q \middle| Q \sum_j P_j = 0\right\}. \end{aligned} \tag{4.3}$$

For the proof, consider the first definition (4.1) of $\mathcal{DP}_C$. First I show that $\mathcal{DP}_C(W)$ is a subset of $\mathcal{F}(\{P_j\}_j)$, and after that I show the converse.

Any projection $\tilde{Q}$ which is a member of $\mathcal{DP}_C(W)$ is a sum of a projection in the Boolean algebra $\mathcal{B}(\{P_j\}_j)$ and a projection in $\mathcal{N}(\{P_j\}_j)$. $\mathcal{B}(\{P_j\}_j)$ is the closure of $\{P_j\}_j$ under $\vee$, $\wedge$, and $\neg$, and is therefore a subset of $\mathcal{F}(\{P_j\}_j)$. $\mathcal{N}(\{P_j\}_j)$ is equal to the set $\{Q | Q \sum_j P_j = 0\}$ and is therefore also a subset of $\mathcal{F}(\{P_j\}_j)$. The projection $\tilde{Q}$ is thus a member of $\mathcal{F}(\{P_j\}_j)$ and $\mathcal{DP}_C(W)$ is thus a subset of $\mathcal{F}(\{P_j\}_j)$.

The set $\mathcal{F}(\{P_j\}_j)$ is the closure under $\vee$, $\wedge$, and $\neg$, of the union of the sets $\{P_j\}_j$ and $\{Q | Q \sum_j P_j = 0\}$. It is easily checked that all members of $\{P_j\}_j$ and all members of $\{Q | Q \sum_j P_j = 0\}$ obey definition (4.1). The union of $\{P_j\}_j$ and $\{Q | Q \sum_j P_j = 0\}$ is thus a subset of $\mathcal{DP}_C(W)$. Clifton proved[8] that $\mathcal{DP}_C(W)$ itself is closed under $\vee$, $\wedge$, and $\neg$, and it therefore follows that $\mathcal{F}(\{P_j\}_j)$ is a subset of $\mathcal{DP}_C(W)$. □

Clifton's property ascription can thus be defined in three different ways. To simplify things, I define in this essay $\mathcal{DP}_C$ as a faux-Boolean algebra, so

$$\mathcal{DP}_C(W) := \mathcal{F}(\{P_j\}_j),$$

and with (4.2) and (4.3) I define a faux-Boolean algebra in its turn as

$$\mathcal{F}(\{P_j\}_j) := \{Q | QP_j = P_j \text{ or } 0, \text{ for any } P_j \text{ in } \{P_j\}_j\}. \tag{4.4}$$

Clifton's property ascription can be understood as follows. Like Kochen, Clifton requires that $\mathcal{DP}$ obey the closure condition. But $\mathcal{DP}$ is not the closure only of $\{P_j\}_j$ but of the union of $\{P_j\}_j$ and two other sets. These two other sets are determined by what Clifton calls the *minimal reality criterion*: "any projection whose value is predictable with certainty in state W should get a definite value."[9] If one now takes "predictable" to mean "predictable as the outcome of a properly calibrated measurement[10] on α," and if one interprets "with certainty" as "with probability 1," this criterion becomes the following condition.

Certainty:
If, given state W, the value of a projection Q is predictable with probability 1 as the outcome of a calibrated measurement of Q on α, then Q has that value.[11]

The projections that have definite values on the basis of this certainty condition are either members of the set $\{Q|Q\,\Sigma_j\,P_j = \Sigma_j\,P_j\}$ or members of $\{Q|Q\,\Sigma_j\,P_j = 0\}$. For the members of $\{Q|Q\,\Sigma_j\,P_j = \Sigma_j\,P_j\}$, the probability $\mathrm{Tr}(QW)$ to obtain a measurement outcome that corresponds to $[Q] = 1$ is equal to 1. For the members of $\{Q|Q\,\Sigma_j\,P_j = 0\}$, the probability $\mathrm{Tr}((\mathbb{1} - Q)W)$ to obtain an outcome that corresponds to $[Q] = 0$ is also equal to 1.

$\mathcal{DP}_C$ can now be constructed starting with the core property ascription and using the certainty and closure conditions. From the core property ascription and the certainty condition it follows that the union of $\{P_j\}_j$, $\{Q|Q\,\Sigma_j\,P_j = 0\}$, and $\{Q|Q\,\Sigma_j\,P_j = \Sigma_j\,P_j\}$ is a subset of $\mathcal{DP}_C$. With the closure condition $\mathcal{DP}_C$ becomes

$$\mathcal{DP}_C(W) = \mathrm{Closure}_{\vee,\wedge,\neg} \text{ of } \{P_j\}_j \cup \left\{Q|Q\sum_j P_j = 0\right\} \cup \left\{Q|Q\sum_j P_j = \sum_j P_j\right\}.$$

This expression for $\mathcal{DP}_C(W)$ is equivalent with (4.3) because $\{Q|Q\,\Sigma_j\,P_j = \Sigma_j\,P_j\}$ is a subset of the closure of $\{P_j\}_j \cup \{Q|Q\,\Sigma_j\,P_j = 0\}$. To prove this, take any element $\tilde{Q}$ which is a member of $\{Q|Q\,\Sigma_j\,P_j = \Sigma_j\,P_j\}$. It holds for the complement $\neg\tilde{Q}$ that $\neg\tilde{Q}\,\Sigma_j\,P_j = (\mathbb{1} - \tilde{Q})\Sigma_j\,P_j = 0$ because $\tilde{Q}\,\Sigma_j\,P_j = \Sigma_j\,P_j$. Therefore $\neg\tilde{Q}$ is in $\{Q|Q\,\Sigma_j\,P_j = 0\}$ and $\tilde{Q}$ is in the closure of $\{P_j\}_j \cup \{Q|Q\,\Sigma_j\,P_j = 0\}$. □

Clifton's proposal can be summarized as

$$\left.\begin{array}{l}\text{Core Property Ascription}\\ \quad+\text{ Certainty}\\ \quad+\text{ Closure}\end{array}\right\} \rightarrow \mathcal{DP}(W) = \mathcal{F}(\{P_j\}_j).$$

5. Beyond Clifton's Property Ascription

My aim is to determine the largest possible set of definite-valued projections without running into contradictions. $\mathcal{DP}_C$ is larger than $\mathcal{DP}_K$, but it still remains to be decided whether consistent property ascriptions even larger than $\mathcal{DP}_C$ exist.

Clifton (1995a) devised an argument to the effect that $\mathcal{DP}$ cannot be larger than $\mathcal{DP}_C$. In outline the argument is as follows.

If one assumes that $\mathcal{DP}(W)$ is a natural property ascription and yields a noncontextual solution of the measurement problem in quantum theory, then $\mathcal{DP}(W)$ should obey six conditions. Four conditions are related to the demand that $\mathcal{DP}(W)$ be a natural property ascription. The first two conditions are the already introduced closure and certainty conditions. The third condition is an invariance condition: because $\mathcal{DP}(W)$ is defined from W, $\mathcal{DP}(W)$ should be invariant under all automorphisms of $\mathcal{H}^\alpha$ that preserve W. Fourth, there is an ignorance interpretation condition: a nonpure density operator W can represent a proper or an improper mixture and W itself does not reveal what kind of mixture it represents; it is thus always possible that W represents a proper mixture, and the value assignment $\mathcal{DP}(W)$ should therefore be consistent with an ignorance interpretation of W. The next conditions are related to the demand that $\mathcal{DP}(W)$ provide a noncontextual solution of the measurement problem. The fifth condition is that the actual value assignments to the projections in $\mathcal{DP}(W)$ be noncontextual and consistent with the predictions of quantum theory. The sixth condition is that $\mathcal{DP}(W)$ contain the eigenprojections of W.

Clifton proves now that $\mathcal{DP}$ obeys these six conditions if and only if $\mathcal{DP} = \mathcal{DP}_{\mathrm{C}}$. (For the precise formulation of the conditions and for further discussion, I refer to Clifton [1995a].)

Alternatively Clifton (1995b) showed that if $\mathcal{DP}$ is supposed to be a local property ascription, $\mathcal{DP}$ should also obey six (partly different) conditions. $\mathcal{DP}$ obeys these conditions if and only if $\mathcal{DP} = \mathcal{DP}_{\mathrm{C}}$. Also Dickson (1995) gave two arguments that, given certain requirements, $\mathcal{DP}_{\mathrm{C}}$ is the only satisfactory property ascription.

One can also give an argument to the effect that $\mathcal{DP}$ might be larger than $\mathcal{DP}_{\mathrm{C}}$. This second line of reasoning makes use of two more conditions one can impose on property ascriptions. These conditions, proposed by Healey (1989) and Arntzenius (1990), are discussed in the next two sections. One can prove that $\mathcal{DP}_{\mathrm{C}}$ does not meet these conditions ($\mathcal{DP}_{\mathrm{K}}$ doesn't meet them either). As a result, if one accepts that property ascriptions should meet these two conditions, $\mathcal{DP}$ cannot be equal to $\mathcal{DP}_{\mathrm{C}}$. This opens up the possibility that $\mathcal{DP}$ is larger than $\mathcal{DP}_{\mathrm{C}}$.

At this point it appears that one has to reject at least one of Clifton's and at least one of Dickson's conditions to define an expanded property ascription that is larger than $\mathcal{DP}_{\mathrm{C}}$. It will turn out, however, that one can obtain an expanded property ascription that is in a quite specific way larger than $\mathcal{DP}_{\mathrm{C}}$ without rejecting their conditions. In section 8 I will develop a new approach to determine the expanded property ascription. With this new approach one can identify two sets of definite-valued projections. The first set, $\mathcal{DP}_{\mathrm{Prob}=1}$, is a set of projections that have definite values with probability 1. The second set, $\mathcal{DP}_{\mathrm{Prob}>0}$, is a set of projections that possibly have definite values, that is, with a probability strictly larger than zero but not necessarily equal to 1. In section 9 it will be shown that this new expanded property ascription meets all the proposed conditions. It is argued that the conditions put forward by Clifton and Dickson should only be imposed on the set $\mathcal{DP}_{\text{Prob}=1}$ and that the conditions proposed by Healey and Arntzenius should only be imposed on the set $\mathcal{DP}_{\text{Prob}>0}$.

To conclude, one can accept the arguments of Clifton and Dickson and at the same time define an expanded property ascription that is different from $\mathcal{DP}_{\text{C}}$. This alternative property ascription is larger than $\mathcal{DP}_{\text{C}}$ in the sense that $\mathcal{DP}_{\text{Prob}>0}$ is larger than $\mathcal{DP}_{\text{C}}$, meets the conditions of Healey and Arntzenius, and goes in this way beyond Clifton's property ascription.

6. Healey's Weakening Condition

Healey (1989) introduced a condition for property ascriptions called the weakening condition.[12] Both $\mathcal{DP}_{\text{K}}$ and $\mathcal{DP}_{\text{C}}$ fail to obey it.

Call a projection Q weaker than another projection P—notation $Q > P$—if the subspace $\mathcal{Q}$ onto which Q projects includes the subspace $\mathcal{P}$ onto which P projects, so if $\mathcal{Q} \supseteq \mathcal{P}$. The weakening condition then reads: "If σ has property P, then σ also has every property Q such that $P < Q$." Formulated in the present notation the condition becomes

Weakening:
If $[P] = 1$, then $\{Q | QP = P\} \subseteq \mathcal{DP}(W)$ with $[Q] = 1$.

Healey motivates this condition with another condition, the property inclusion condition. This last condition states that the proposition "the value of A is in interval Δ" is true only if the proposition "the value of A is in interval Γ" is true for every $\Delta \subseteq \Gamma$. If one adopts the weakening condition, the property inclusion condition is fulfilled. Take for instance a magnitude A with spectral resolution $\Sigma_{j=1} a_j |a_j\rangle\langle a_j|$ where $a_j = j$, and consider a system for which $[|a_1\rangle\langle a_1|] = 1$. This value assignment implies that it is true that "the value of A is in interval $[\frac{1}{2}, 1\frac{1}{2}]$." If property inclusion holds, it should also be true that "the value of A is in $[\frac{1}{2}, 1\frac{1}{2} + n]$" with $n \in \mathbb{N}$. These last truth-valuations imply according to Healey that $[\Sigma_{j=1}^{n+1} |a_j\rangle\langle a_j|] = 1$ for all $n \in \mathbb{N}$. The weakening condition guarantees that this implication holds: because all projections $\Sigma_{j=1}^{n+1} |a_j\rangle\langle a_j|$ are weaker than $|a_1\rangle\langle a_1|$, these projections have value 1.

To prove that $\mathcal{DP}_{\text{K}}$ and $\mathcal{DP}_{\text{C}}$ in general do not satisfy the weakening condition, take a system α with a state with spectral resolution

$$W = \tfrac{1}{2}|w_1\rangle\langle w_1| + \tfrac{1}{3}|w_2\rangle\langle w_2| + \tfrac{1}{6}|w_3\rangle\langle w_3|. \quad (6.1)$$

Assume now that $[|w_1\rangle\langle w_1|] = 1$. Then with the weakening condition the projection

$$Q = |w_1\rangle\langle w_1| + \tfrac{1}{2}(|w_2\rangle + |w_3\rangle)(\langle w_2| + \langle w_3|)$$

should be a member of $\mathcal{DP}(W)$. But Q is a member neither of $\mathcal{B}(\{|w_j\rangle\langle w_j|\}_{j=1}^{3})$ nor of $\mathcal{F}(\{|w_j\rangle\langle w_j|\}_{j=1}^{3})$ and is thus a member neither of $\mathcal{DP}_{\mathrm{K}}(W)$ nor of $\mathcal{DP}_{\mathrm{C}}(W)$.

7. Arntzenius's Criticism

Arntzenius (1990) criticized Kochen's property ascription by arguing that it violates an intuition about the outcomes of measurements.[13] Clifton's property ascription also violates this intuition.

Consider a measurement device μ with pointer positions (possible measurement outcomes) represented by orthogonal projections $\{M_j^\mu\}_j$. (The superscript μ indicates that these projections are defined in the Hilbert space $\mathcal{H}^\mu$ associated with the device μ.) The intuition is now that if, after a measurement, it is true that the device possesses a specific pointer position M_k^μ, then it should be false that the device possesses another pointer position M_j^μ, for each $j \neq k$. This intuition is present in everyday reasonings, such as "because it is true that this fish weighs 4 kilos, it is false that this fish weighs X kilos" where X can be any number different from 4. Let's capture the intuition by means of a condition for property ascriptions:

Arntzenius:
For a measurement device μ with pointer positions represented by orthogonal projections $\{M_j^\mu\}_j$, it must hold after a measurement that

$$\text{If } [M_k^\mu] = 1, \text{ then } \{M_j^\mu\}_{j \neq k} \subseteq \mathcal{DP}(W^\mu) \text{ with } [M_j^\mu] = 0.$$

$\mathcal{DP}_K$ and $\mathcal{DP}_C$ do not satisfy this condition. Take a magnitude $A^\alpha = \Sigma_{j=1}\, a_j|a_j^\alpha\rangle\langle a_j^\alpha|$ pertaining to a system α and measure this magnitude by means of a von Neumann measurement. Consider an interaction between α and a measurement device μ such that an initial state $\Sigma_{j=1} c_j|a_j^\alpha\rangle|m_0^\mu\rangle$ of the composite of α and μ evolves to the final state $\Sigma_{j=1}\, c_j|a_j^\alpha\rangle|m_j^\mu\rangle$. Here, the projection $|m_0^\mu\rangle\langle m_0^\mu|$ represents the ready-to-measure state of μ before the measurement and the orthogonal projections $\{|m_j^\mu\rangle\langle m_j^\mu|\}_{j=1}$ represent the pointer positions $\{M_j^\mu\}_j$. The final state of the device is $W^\mu = \Sigma_{j=1}\, |c_j|^2|m_j^\mu\rangle\langle m_j^\mu|$.

Suppose now that $|c_4|^2 = \frac{1}{2}$, and $|c_5|^2 = |c_6|^2 = \frac{1}{4}$, and that all other $|c_j|^2$ are equal to zero. The final state of μ is then

$$W^\mu = \tfrac{1}{2}|m_4^\mu\rangle\langle m_4^\mu| + \tfrac{1}{4}(|m_5^\mu\rangle\langle m_5^\mu| + |m_6^\mu\rangle\langle m_6^\mu|).$$

The properties possessed by μ are then according to Kochen and Clifton respectively

$$\mathcal{DP}_K(W^\mu) = \mathcal{B}(|m_4^\mu\rangle\langle m_4^\mu|, |m_5^\mu\rangle\langle m_5^\mu| + |m_6^\mu\rangle\langle \mathrm{m}_6^\mu|)$$
$$\mathcal{DP}_C(W^\mu) = \mathcal{F}(|m_4^\mu\rangle\langle m_4^\mu|, |m_5^\mu\rangle\langle m_5^\mu| + |m_6^\mu\rangle\langle m_6^\mu|).$$

The core property ascription assigns probability $\frac{1}{2}$ to the possibility that $|m_4^\mu\rangle\langle m_4^\mu|$ has value 1. The antecedent of the Arntzenius condition can thus be met—but the consequent can never be met. If the pointer positions $\{|m_j^\mu\rangle\langle m_j^\mu|\}_{j\neq 4}$ all must have value 0, then the projections $\{|m_j^\mu\rangle\langle m_j^\mu|\}_{j\neq 4}$ must have definite values and thus have to be members of $\mathcal{DP}_K(W^\mu)$ or $\mathcal{DP}_C(W^\mu)$, respectively. Yet this is not the case: not a single projection in $\{|m_j^\mu\rangle\langle m_j^\mu|\}_{j\neq 4}$ is a member of $\mathcal{DP}_K(W^\mu)$ and $|m_5^\mu\rangle\langle m_5^\mu|$ and $|m_6^\mu\rangle\langle m_6^\mu|$ are not members of $\mathcal{DP}_C(W^\mu)$.

As a consequence, what can happen in Kochen's and Clifton's property ascription is that it is true that a specific fish weighs 4 kilos ($[|m_4^\mu\rangle\langle m_4^\mu|] = 1$), but that it is neither true nor false that that same fish weighs 5 kilos ($|m_5^\mu\rangle\langle m_5^\mu|$ has no definite value).

8. The New Expanded Property Ascription

In order to develop a new expanded property ascription, consider again the core property ascription. It has two aspects. First, it defines a set of definite-valued projections $\{P_j\}_j$. Second, it gives specific value assignments: with probability $\mathrm{Tr}(P_k W)$, $[P_k] = 1$ and $[P_j] = 0$ for all $j \neq k$. In his determination of $\mathcal{DP}_C(W)$, Clifton uses the first aspect but ignores the second. As a result $\mathcal{DP}_C(W)$ is independent of the specific value assignment: the possibility that if, for instance, $[P_1] = 1$ other projections are definite-valued than if $[P_2] = 1$, is disregarded.

This independence has its advantages—members of $\mathcal{DP}_C$ always have definite values regardless of the specific value assignment—but has its disadvantages, too. The weakening condition and the Arntzenius condition refer to specific value assignments, and $\mathcal{DP}_C(W)$ fails to obey them. The weakening condition, for instance, requires that if P has value 1, then the members of the set $\{Q|QP = P\}$ also have definite values. But if P doesn't have value 1, the weakening condition requires nothing. So, if $[P] = 0$, members of $\{Q|QP = P\}$ need not have definite values. A property ascription can thus obey the weakening condition by assigning values to projections conditional on whether $[P] = 1$ or not.

This conditionalization on the specific value assignment is an essential point in the new expanded property ascription. I start with the first aspect of the core property ascription, so $\{P_j\}_j \subseteq \mathcal{DP}(W)$. Following Kochen, I adopt the closure condition, and following Clifton, I also accept the condition that any projection Q whose value is predictable with probability 1 as the outcome of a measurement should get that value. But in order to incorporate the second aspect of the core property ascription, that is, the specific value assignment, I modify the certainty condition. In the new condition, called conditional certainty, the predictions about values of projections Q are conditionalized on the specific value assignment to $\{P_j\}_j$:

Conditional Certainty:
If, given state W and given assignment $[P_a] = 1$, the value of a projection Q is predictable with probability 1 as the outcome of a calibrated measurement of Q on α, then Q has that value.

In order to apply this new condition, one has to determine with which probabilities one can predict outcomes of measurements of Q. In the case of the certainty condition, one could use the expectation value $\text{Tr}(QW)$ to determine these probabilities (see section 4). In the present case, expectation values don't deliver the goods because they yield probabilities given W, and not given W *and* given $[P_a] = 1$. A more appropriate way to determine the desired probabilities is by means of transition probabilities between the value assignment $[P_a] = 1$ and the presence of an outcome $[M_b^\mu] = 1$ after the measurement of Q (here, M_b^μ represents a pointer position of a measurement device μ). A precondition of this way of determining the desired probabilities is that after the measurement of Q the pointer positions of the measurement device are assigned definite values.

Let's choose now Lüders measurements[14] as the calibrated measurements. Vermaas (1996) shows that after a Lüders measurement the pointer positions in general do have definite values. Further, it is shown in the appendix of this chapter that in the case of a Lüders measurement of projection Q on a system α with state W and $[P_a] = 1$, the predicted probabilities for outcomes corresponding to $[Q] = 1$ and $[Q] = 0$ are given by

$$\begin{aligned} \text{Prob}([Q] = 1/[P_a] = 1) &= \text{Tr}(QP_a)/\text{Tr}(P_a) \\ \text{Prob}([Q] = 0/[P_a] = 1) &= \text{Tr}((\mathbb{1} - Q)P_a)/\text{Tr}(P_a) \end{aligned}$$

provided that $\text{Tr}(QP_a) \neq \frac{1}{2}$.

Let's determine which projections get definite values with these formulas for outcomes of Lüders measurements. Take a system α with state W and assume that $[P_a] = 1$. Let $\mathcal{DP}(W,[P_a] = 1)$ be the set of projections that get definite values according to the new expanded property ascription. In the appendix it is proved that

$$\mathcal{DP}(W,[P_a] = 1) = \mathcal{F}(P_a) \tag{8.1}$$

and that the values of the projections in $\mathcal{DP}(W,[P_a] = 1)$ are given by

$$[Q] = \mathrm{Tr}(QP_a)/\mathrm{Tr}(P_a). \tag{8.2}$$

These results hold if one takes Lüders measurements when applying the conditional certainty condition. The question is now whether these results also hold if one takes other calibrated measurements. I discuss this question only tentatively, primarily because it is not yet possible to answer it fully.

Consider an arbitrary calibrated measurement of a projection Q. Take a measurement device μ and let it interact with α such that $\mathrm{Tr}^{\mu}(M_1^{\mu}W^{\mu})$ is equal to $\mathrm{Tr}(QW)$; W^{μ} is the state μ after the measurement and M_1^{μ} is the pointer position that corresponds to Q. A first problem, pointed out by Albert and Loewer (1990), is that if one applies the modal interpretation to the state W^{μ} of μ after the measurement, it can happen that the pointer positions $\{M_j^{\mu}\}_j$ don't get definite values. This violates the precondition of the way in which the probabilities for the presence of the measurement outcomes are determined. Bacciagaluppi and Hemmo (1996) addressed this first problem. They showed by means of decoherence that after the measurement the modal interpretation assigns definite value to projections that are very close to the pointer positions. Let's assume that the results of Bacciagaluppi and Hemmo solve this first problem sufficiently. A second problem is then that there don't yet exist general solutions for transition probabilities in the modal interpretation between the initial value assignment $[P_a] = 1$ and the final value assignments to the pointer positions (Vermaas 1996).

It is thus not yet possible to determine which projections get definite values with the conditional certainty condition if an arbitrary calibrated measurement is considered. But insofar as the transition probabilities between an initial $[P_a] = 1$ and the final measurement outcomes can be determined—for instance if the state of α is not entangled with its environment before the measurement—the results (8.1) and (8.2) found by means of Lüders measurements are vindicated.

Let's call the property ascription (8.1) and (8.2) the *conditional property ascription.* This conditional property ascription cannot simply be compared with Clifton's property ascription. The set $\mathcal{DP}_{\mathrm{C}}(W)$ is a set of definite-valued projections given a state W; the set $\mathcal{DP}(W,[P_a] = 1)$ is a set of definite-valued projections given a state W *and* given the assignment $[P_a] = 1$. In order to compare the new property ascription with Clifton's property ascription, one therefore has to get rid of the conditionalization on $[P_a] = 1$. This can be done in two ways. First, one can define the set of projections which possibly (with a probability > 0) have definite values given a state W. Second, one can define the set of projections that always (with a probability equal to 1) have definite values given a state W. This first set is the union of all conditional property ascriptions $\mathcal{DP}(W,[P_1] = 1)$, $\mathcal{DP}(W,[P_2] = 1)$, and so on; the second set is the intersection of all these conditional property ascriptions. Let $\mathcal{DP}_{\text{Prob}>0}(W)$ signify the set of possible definite-valued projections and let $\mathcal{DP}_{\text{Prob}=1}(W)$ signify the set of projections that are always definite-valued. The set $\mathcal{DP}_{\text{Prob}>0}(W)$ is equal to

$$\mathcal{DP}_{\text{Prob}>0}(W) = \cup_j \mathcal{F}(P_j). \tag{8.3}$$

The probabilities with which the members of this set have definite values, and their specific values, are generated by the core property ascription and the conditional property ascription (8.1) and (8.2):

$$\begin{array}{l}\text{With probability } \mathrm{Tr}(P_jW) \text{ the value of } Q \in \mathcal{F}(P_j) \text{ is} \\ \quad [Q] = \mathrm{Tr}(QP_j)/\mathrm{Tr}(P_j), \\ \text{and projections } \tilde{Q} \notin \mathcal{F}(P_j) \text{ do not have values.}\end{array} \tag{8.4}$$

The set $\mathcal{DP}_{\text{Prob}=1}(W)$ is equal to $\cap_j \mathcal{F}(P_j)$. Inspection of the definition (4.4) for faux-Boolean algebras yields that $\cap_j \mathcal{F}(P_j) = \mathcal{F}(\{P_j\}_j)$, so

$$\mathcal{DP}_{\text{Prob}=1}(W) = \mathcal{F}(\{P_j\}_j). \tag{8.5}$$

The specific values of the members of this set are given by:

$$\begin{array}{l}\text{With probability } \mathrm{Tr}(P_jW) \text{ the value of } Q \text{ is} \\ \quad [Q] = \mathrm{Tr}(QP_j)/\mathrm{Tr}(P_j).\end{array} \tag{8.6}$$

The sets $\mathcal{DP}_{\text{Prob}>0}(W)$ and $\mathcal{DP}_{\text{Prob}=1}(W)$ can be compared with Clifton's property ascription. The conclusion of such a comparison is that on the one hand, given a state W, many more projections can possibly have definite value with the new property ascription than with Clifton's property ascription:

$$\mathcal{DP}_{\text{C}}(W) \subseteq \mathcal{DP}_{\text{Prob}>0}(W).$$

On the other hand the set of projections that always have definite values, given a state W, are in both approaches equal:

$$\mathcal{DP}_{\text{C}}(W) = \mathcal{DP}_{\text{Prob}=1}(W).$$

The new expanded property ascription can thus be seen as an extension of Clifton's property ascription: the set of projections that always have definite values is equal in both property ascriptions, but the new property ascription adds a set of projections that possibly have definite values.

The new expanded property ascription can be summarized as

$$\left.\begin{array}{l}\text{Core Property Ascription}\\ \quad+\text{ Conditional Certainty}\\ \quad+\text{ Closure}\end{array}\right\} \to \mathcal{DP}(W,[P_a]=1)=\mathcal{F}(P_a),$$

$$\to \left[\begin{array}{rcl}\mathcal{DP}_{\text{Prob}>0}(W) &=& \cup_j \mathcal{F}(P_j)\\ \mathcal{DP}_{\text{Prob}=1}(W) &=& \mathcal{F}(\{P_j\}_j).\end{array}\right.$$

9. Evaluation of the New Property Ascription

Because of its dual characterization—$\mathcal{DP}_{\text{Prob}>0}$ and $\mathcal{DP}_{\text{Prob}=1}$—the new property ascription satisfies on the one hand the weakening condition and the Arntzenius condition, and endorses on the other hand Clifton's and Dickson's conclusions.

Consider first the weakening condition. As stated above, a property ascription satisfies this condition if, given that $[P]=1$, all projections Q in $\{Q|QP=P\}$ have value 1. The members of $\{Q|QP=P\}$ need not *always* have definite values. It is thus unnecessarily strict to require that the set $\mathcal{DP}_{\text{Prob}=1}$ of the new property ascription obey the weakening condition; it suffices to check whether $\mathcal{DP}_{\text{Prob}>0}$ meets this condition.

According to the value assignment (8.4) to $\mathcal{DP}_{\text{Prob}>0}$, the projections that have simultaneously definite values are all member of $\mathcal{F}(P_j)$ where P_j is a certain eigenprojection of W. If the antecedent of the weakening condition holds—that is, if $[P]=1$—then this eigenprojection P_j necessarily obeys the relation $PP_j=P_j$. Proof: Because P has a definite value, P is a member of $\mathcal{F}(P_j)$. With the definition (4.4) of $\mathcal{F}(P_j)$ this is the case only if $PP_j=P_j$ or $PP_j=0$. If $PP_j=0$, it follows with the value assignment (8.4) to $\mathcal{DP}_{\text{Prob}>0}$ that $[P]=0$. Because $[P]=1$, the conclusion is that $PP_j=P_j$. $\square$

From the relation $PP_j=P_j$ one can show that if $[P]=1$, the consequent of the weakening condition also holds. Take any Q with $QP=P$. Such a Q obeys

$$QP_j = QPP_j = PP_j = P_j \tag{9.1}$$

(use the relation $PP_j=P_j$ twice and use $QP=P$ once). From this identity (9.1) it follows that Q is a member of $\mathcal{F}(P_j)$, which implies that Q has a definite value simultaneously with P. The specific value of Q is with (9.1) and the value assignment (8.4) to $\mathscr{DP}_{\text{Prob}>0}$, equal to 1. The new property ascription thus meets the weakening condition via $\mathscr{DP}_{\text{Prob}>0}$.

Consider second the Arntzenius condition. It can be proved that the new property ascription satisfies the following stronger condition, which applies to any system α, not only to measurement devices:

Arntzenius*:
If $[P] = 1$, then $\{Q|QP = 0\} \subseteq \mathscr{DP}(W)$ with $[Q] = 0$.

Clearly this condition is stronger than the original Arntzenius condition: if the Arntzenius* condition holds and α is a measurement device μ and $[M_k^\mu] = 1$ after a measurement, then all the other pointer positions $\{M_j^\mu\}_{j \neq k}$ have value 0 because they are in the set $\{Q^\mu|Q^\mu M_k^\mu = 0\}$.

The Arntzenius* condition requires that the projections in the set $\{Q|QP = 0\}$ have definite value 0 only if $[P] = 1$. It is therefore again unnecessarily strict to require that the set $\mathscr{DP}_{\text{Prob}=1}$ obey this condition, so consider $\mathscr{DP}_{\text{Prob}>0}$.

The projections that have simultaneous definite values are according to (8.4) members of $\mathscr{F}(P_j)$ with P_j a certain eigenprojection of W. If the antecedent of the Arntzenius* condition holds—that is, if $[P] = 1$—it follows as shown above that $PP_j = P_j$. Consider next the consequent of the Arntzenius* condition. Take any Q with $QP = 0$. Such a Q obeys

$$QP_j = QPP_j = 0 \tag{9.2}$$

and so is a member of $\mathscr{F}(P_j)$ and thus has a definite value simultaneously with P. The specific value of Q is with (9.2) and the value assignment (8.4) to $\mathscr{DP}_{\text{Prob}>0}$, equal to 0. The new property ascription thus meets also the Arntzenius* condition via $\mathscr{DP}_{\text{Prob}>0}$.

Consider third Clifton's argument, described in section 5, that $\mathscr{DP}$ is not larger than $\mathscr{DP}_C$. In order to determine whether the new expanded property ascription confirms that conclusion, it first has to be decided to which set of definite-valued projections the argument refers: to $\mathscr{DP}_{\text{Prob}>0}$, to $\mathscr{DP}_{\text{Prob}=1}$, or to both.

Clifton's argument is that if $\mathscr{DP}$ is supposed to be a natural property ascription which yields a noncontextual solution of the measurement problem, $\mathscr{DP}$ must obey six conditions. One of these conditions is the closure condition, which reveals that it is a tacit assumption in the argument that all members of $\mathscr{DP}$ have a definite value with probability 1.[15] To see this, consider Clifton's motivation why $\mathscr{DP}$ needs to obey the closure condition. Clifton says that it would be preferred that the members of $\mathscr{DP}$ formed a logic of propositions within which conjunctions, disjunctions, and negations can be formed.[16] Given this assumption that $\mathscr{DP}$ forms a logic, one indeed can argue that $\mathscr{DP}_{\text{Prob}=1}$ needs to obey the closure condition. But given this assumption, one cannot conclude that $\mathcal{DP}_{\text{Prob}>0}$ needs to obey closure.

Take two projections P and Q, both members of $\mathcal{DP}_{\text{Prob}=1}$. Then the propositions "α possesses P" and "α possesses Q" are both with probability 1 true or false, and given that $\mathcal{DP}_{\text{Prob}=1}$ forms a logic, the conjunction of these propositions must also be true or false with probability 1. Hence, $P \wedge Q$ must have a definite value with probability 1. $P \wedge Q$ must thus be a member of $\mathcal{DP}_{\text{Prob}=1}$, and one can conclude that $\mathcal{DP}_{\text{Prob}=1}$ needs to be closed under the meet $\wedge$. One can run the same argument for the join $\vee$ and the orthocomplement $\neg$.

Take now two projections P and Q, both members of $\mathcal{DP}_{\text{Prob}>0}$. If there is probability > 0 that P and Q have *simultaneously* definite values, the above reasoning is still valid. With probability > 0 the propositions "α possesses P" and "α possesses Q" have then simultaneously truth-values, and given that $\mathcal{DP}_{\text{Prob}>0}$ forms a logic, the conjunction of these propositions must also have a truth-value with probability > 0. Hence, $P \wedge Q$ must have a definite value with probability > 0 and $P \wedge Q$ thus needs to be a member of $\mathcal{DP}_{\text{Prob}>0}$. But if there is probability 0 that P and Q have simultaneously definite values (and that is indeed possible for members of $\mathcal{DP}_{\text{Prob}>0}$), the above reasoning no longer applies. For if "α possesses P" and "α possesses Q" have with probability 0 simultaneously truth-values, one can only demand that the conjunction of these propositions have a truth-value with probability 0. As a result, $P \wedge Q$ needs to have a definite value with probability 0, and it doesn't follow that $P \wedge Q$ is a member of $\mathcal{DP}_{\text{Prob}>0}$. If P and Q are members of $\mathcal{DP}_{\text{Prob}>0}$ but have with probability 0 simultaneously definite values, one cannot conclude that because $\mathcal{DP}_{\text{Prob}>0}$ is a logic, $P \wedge Q$ needs to be a member of $\mathcal{DP}_{\text{Prob}>0}$. Hence, $\mathcal{DP}_{\text{Prob}>0}$ does not need to be closed under the meet $\wedge$ and hence $\mathcal{DP}_{\text{Prob}>0}$ does not need to obey the closure condition. Instead, one can only demand with the above reasoning that subsets of projections which have simultaneously definite values obey the closure condition.[17]

Having restricted Clifton's argument to the set $\mathcal{DP}_{\text{Prob}=1}$, one can check whether the new property ascription confirms the conclusion of the argument. The new property ascription is indeed supposed to be a natural property ascription yielding a noncontextual solution of the measurement problem. Hence, $\mathcal{DP}_{\text{Prob}=1}$ must obey the six conditions and $\mathcal{DP}_{\text{Prob}=1}$ thus must be equal to $\mathcal{DP}_{\text{C}}$. As already observed in equation (8.7), $\mathcal{DP}_{\text{Prob}=1}$ is indeed equal to $\mathcal{DP}_{\text{C}}$, so the new property ascription endorses the conclusion of Clifton's argument.

With respect to the other arguments that $\mathcal{DP}$ is equal to $\mathcal{DP}_{\text{C}}$, given by Clifton (1995b) and Dickson (1995), the same conclusion holds. All these arguments assume that $\mathcal{DP}$ should obey the closure condition. The arguments refer therefore to $\mathcal{DP}_{\text{Prob}=1}$, and their conclusion that $\mathcal{DP}_{\text{Prob}=1}$ is equal to $\mathcal{DP}_{\text{C}}$ is confirmed by the new expanded property ascription.

10. Definite-valued Magnitudes

A value assignment to projections in the sets $\mathcal{DP}_{\text{Prob}>0}$ and $\mathcal{DP}_{\text{Prob}=1}$ induces a value assignment to magnitudes represented by arbitrary operators.

Take a physical magnitude represented by a self-adjoint, hypermaximal operator A defined in $\mathcal{H}^{\alpha}$. Let $\{\Delta\}$ represent the Borel sets on the real line $\mathbb{R}$ and let $\{E_A(\Delta)\}_\Delta$ be the projections that form the spectral family of A such that

$$A = \int_{\lambda=-\infty}^{\infty} \lambda dE_A((-\infty, \lambda]).$$

The spectral family $\{E_A(\Delta)\}_\Delta$ has the properties:

$$\begin{aligned} E_A(\emptyset) &= 0 \\ E_A(\mathbb{R}) &= \mathbb{1} \\ E_A(\Delta \cap \Gamma) &= E_A(\Delta)E_A(\Gamma) \\ E_A(\Delta \cup \Gamma) &= E_A(\Delta) + E_A(\Gamma) - E_A(\Delta)E_A(\Gamma) \\ E_A(\text{IR} - \Delta) &= \mathbb{1} - E_A(\Delta). \end{aligned} \tag{10.1}$$

Being projections, the members of the spectral family can have definite values. If a member $E_A(\Delta)$ has value 1, one can assign a set of values to the magnitude A. Let $[A]$ be the value of A. The value assignment to A is then given through the following equivalence: $[E_A(\Delta)] = 1$ if and only if $[A]$ is restricted to Δ.

Let the notation for this value assignment be $[A] \in^* \Delta$. The motivation behind this is that on the one hand the value assignment $[A] \in^* \Delta$ has many properties in common with the set theoretic proposition $[A] \in \Delta$. But on the other hand $[A] \in^* \Delta$ differs on one important point from $[A] \in \Delta$: as will be shown, one cannot in general interpret $[A] \in^* \Delta$ as "$[A]$ has a precise value x and x is an element of Δ." Strictly speaking, the "value assignment" $[A] \in^* \Delta$ is therefore not an assignment of a value to A but an assignment of a set of values.

If $E_A(\Delta)$ has value 0, one can also assign a set of values to A. I consider $[E_A(\Delta)] = 0$ as the negation of $[E_A(\Delta)] = 1$, so as equivalent with "not $[A] \in^* \Delta$." The negation in this latter statement can be taken as an exclusion negation or as a choice negation. Accordingly, the statement implies "$[A] \in^* \Gamma$ and $\Gamma \neq \Delta$" or "$[A] \in^* \mathbb{R} - \Delta$." I adopt here the choice negation:[18] $[E_A(\Delta)] = 0$ if and only if $[A] \in^* \mathbb{R} - \Delta$. Let the notation for this value assignment be $[A] \notin^* \Delta$.

From the properties (10.1) of the spectral family and from the new expanded property ascription it follows that the value assignment to A has the following properties:

$$\begin{aligned} &[A] \in^* \mathbb{R} \\ &[A] \notin^* \emptyset \\ &[A] \in^* \Delta \Rightarrow [A] \notin^* \mathbb{R} - \Delta \\ &[A] \notin^* \Delta \Rightarrow [A] \in^* \mathbb{R} - \Delta. \end{aligned} \tag{10.2}$$

The first two properties are easily proved: $\mathbb{1}$ and 0 are both members of $\mathcal{DP}_{\text{Prob}=1}(W)$, and with probability 1 it is the case that $[\mathbb{1}] = 1$ and $[0] = 0$. Hence, with the properties (10.1) it follows that $[A] \in^* \mathbb{R}$ and $[A] \notin^* \emptyset$ with probability 1 . The third property is equivalent with

$$[E_A(\Delta)] = 1 \Rightarrow [E_A(\mathbb{R} - \Delta)] = 0$$

and can be proved with the Arntzenius* condition because $E_A(\mathbb{R} - \Delta)E_A(\Delta) = 0$. The fourth property is equivalent with the requirement

$$[E_A(\Delta)] = 0 \Rightarrow [E_A(\mathbb{R} - \Delta)] = 1.$$

This requirement is conditional on the value assignment $[E_A(\Delta)] = 0$, so should be proved using $\mathcal{DP}_{\text{Prob}>0}$. Assume that $[E_A(\Delta)] = 0$. According to the value assignment (8.4) to $\mathcal{DP}_{\text{Prob}>0}(W)$, this can only be the case if an eigenprojection P_j of W exists such that $[P_j] = 1$, $E_A(\Delta) \in \mathcal{F}(P_j)$, and $E_A(\Delta)P_j = 0$. As a consequence, all projections with definite values simultaneously with $[E_A(\Delta)] = 0$ are also members of $\mathcal{F}(P_j)$. A faux-Boolean algebra $\mathcal{F}(P_j)$ is closed under $\neg$, so $\neg E_A(\Delta) = \mathbb{1} - E_A(\Delta)$ is also a member of $\mathcal{F}(P_j)$. Because $(\mathbb{1} - E_A(\Delta))P_j = P_j - E_A(\Delta)P_j = P_j$, the value of $\mathbb{1} - E_A(\Delta)$ is on the basis of (8.4) equal to 1. According to the properties of the spectral family, $\mathbb{1} - E_A(\Delta)$ is equal to $E_A(\mathbb{R} - \Delta)$. Hence $[E_A(\mathbb{R} - \Delta)] = 1$. □

A second series of properties of the value assignment to A is:

$$\begin{aligned} &[A] \in^* \Delta \Rightarrow [A] \in^* \Gamma \text{ for every } \Gamma \text{ with } \Delta \subseteq \Gamma \\ &[A] \in^* \Delta \Rightarrow [A] \notin^* \Gamma \text{ for every } \Gamma \text{ with } \Delta \cap \Gamma = \emptyset \\ &[A] \notin^* \Delta \Rightarrow [A] \notin^* \Gamma \text{ for every } \Gamma \text{ with } \Gamma \subseteq \Delta \\ &[A] \notin^* \Delta \Rightarrow [A] \in^* \Gamma \text{ for every } \Gamma \text{ with } \Delta \cup \Gamma = \mathbb{R}. \end{aligned} \tag{10.3}$$

The first property is the already mentioned property inclusion condition introduced by Healey (see section 6). The proof is as follows. Given that

$[A] \in^* \Delta$, the projection $E_A(\Delta)$ has value 1. With the weakening condition all projections Q with $QE_A(\Delta) = E_A(\Delta)$ also have value 1. With (10.1) it follows that

$$E_A(\Gamma)E_A(\Delta) = E_A(\Delta \cap \Gamma) = E_A(\Delta)$$

for every Γ with $\Delta \subseteq \Gamma$. Hence $E_A(\Gamma)$ has value 1 and $[A] \in^* \Gamma$. The second property of (10.3) can be proved with the Arntzenius* condition, and the third and fourth follow with the first and second together with the properties (10.2).

A final series of properties concerns the propositional logic of the value assignment:

$$\begin{array}{rcl} [A] \in^* \Delta \quad \text{and} \quad [A] \in^* \Gamma & \Leftrightarrow & [A] \in^* \Delta \cap \Gamma \\ [A] \in^* \Delta \quad \text{or} \quad [A] \in^* \Gamma & \Rightarrow & [A] \in^* \Delta \cup \Gamma \\ \text{not } [A] \in^* \Delta \ (:= [A] \notin^* \Delta) & \Leftrightarrow & [A] \in^* \mathbb{R} - \Delta. \end{array} \tag{10.4}$$

I give the proof of the first relation; the proofs of the second and third relation follow from the properties (10.2) and (10.3).

The first relation is equivalent with

$$[E_A(\Delta)] = 1 \quad \text{and} \quad [E_A(\Gamma)] = 1 \quad \Leftrightarrow \quad [E_A(\Delta \cap \Gamma)] = 1.$$

Both the sufficiency and the necessity part of this relation are conditional on specific value assignments, so the relation should be proved using $\mathcal{DP}_{\text{Prob}>0}$. Necessity follows with the weakening condition, so only sufficiency has to be proved.

Assume that simultaneously $[E_A(\Delta)] = 1$ and $[E_A(\Gamma)] = 1$. According to the value assignment (8.4) to $\mathcal{DP}_{\text{Prob}>0}(W)$, this can only be the case if there exists an eigenprojection P_j of W such that $[P_j] = 1$, $E_A(\Delta)P_j = P_j$, and $E_A(\Gamma)P_j = P_j$. As a consequence all projections $\{Q\}$ with definite values simultaneously with $[E_A(\Delta)] = 1$ and $[E_A(\Gamma)] = 1$ are members of the faux-Boolean algebra $\mathcal{F}(P_j)$ and the values of these projections are $[Q] = \text{Tr}(QP_j)/\text{Tr}(P_j)$. The projection $E_A(\Delta \cap \Gamma)$ is with (10.1) equal to $E_A(\Delta)E_A(\Gamma)$. With the relations $E_A(\Delta)P_j = P_j$ and $E_A(\Gamma)P_j = P_j$, it follows that $E_A(\Delta \cap \Gamma)P_j = P_j$. The projection $E_A(\Delta \cap \Gamma)$ is thus a member of $\mathcal{F}(P_j)$ and has definite value 1. □

One relation is missing in the list (10.4):

$$[A] \in^* \Delta \cup \Gamma \quad \Rightarrow \quad [A] \in^* \Delta \quad \text{or} \quad [A] \in^* \Gamma. \tag{10.5}$$

This relation does not hold in general. Consider, for instance, a system with state

$$W = E_A((1,3])/\text{Tr}(E_A((1,3])).$$

The sets $\mathcal{DP}_{\text{Prob}>0}(W)$ and $\mathcal{DP}_{\text{Prob}=1}(W)$ are then both equal to $\mathcal{F}(E_A((1,3]))$. The projections $E_A((1,2])$ and $E_A((2,3])$ are not members of $\mathcal{F}(E_A((1,3]))$ and thus do not have definite values. It follows that

$$[A] \in^* (1,3] \quad \not\Rightarrow \quad [A] \in^* (1,2] \quad \text{or} \quad [A] \in^* (2,3]. \qquad (10.6)$$

This example proves too that $[A] \in^* \Delta$ cannot be taken as "$[A]$ has a precise value x and x is an element of Δ." For if $[A] \in^* (1,3]$ implies that $[A]$ has a precise value x and x is an element of $(1,3]$, then x has to be an element of either $(1,2]$ or $(2,3]$. It would then hold that either $[A] \in^* (1,2]$ or $[A] \in^* (2,3]$. But given (10.6), this need not be the case.

If a magnitude is represented by an operator A with a discrete spectrum,

$$A = \sum_j a_j Q_j$$

(eigenprojections corresponding with an eigenvalue equal to zero are part of this sum), the members of the spectral family become

$$E_A(\Delta) = \sum_{a_j \in \Delta} Q_j.$$

In the special case that all eigenprojections $\{Q_j\}_j$ have simultaneously definite values—if an eigenprojection P_j of W exists such that $\{Q_j\}_j \subseteq \mathcal{F}(P_j)$ and $[P_j] = 1$ or, even stronger, if $\{Q_j\}_j \subseteq \mathcal{DP}_{\text{Prob}=1}(W)$—then the value assignment to A yields precise values. If Q_k has value 1, the projection $E_A(\{a_k\}) = Q_k$ has also value 1 and thus $[A] = a_k$. Given the second relation of (10.3) all the projections $E_A(\{x\})$ with $x \neq a_k$ have then value 0, hence $[A] \neq x$ if $x \neq a_k$. If Q_k has value 0, it follows that $[A] \neq a_k$.

In this special case in which A has precise values, the missing relation (10.5) does hold.

11. Conclusions

In this chapter a new expanded property ascription in the modal interpretation has been developed. For a system α with a state given by a density operator W, this property ascription defines two sets of projections with definite values. The first set, $\mathcal{DP}_{\text{Prob}>0}(W)$, contains projections that possibly (with probability > 0) have definite values. The second set, $\mathcal{DP}_{\text{Prob}=1}(W)$, contains projections that always (with probability equal to 1) have definite values.

These two sets are defined in terms of faux-Boolean algebras generated by eigenprojections $\{P_j\}_j$ of the state W that correspond to nonzero eigenvalues. A faux-Boolean algebra $\mathcal{F}(\{S_j\}_j)$ generated by mutual orthogonal projections $\{S_j\}_j$ is defined as the set of projections $\{Q\}$ for which holds

$$\mathcal{F}(\{S_j\}_j) := \{Q \mid QS_j = S_j \text{ or } 0, \text{ for any } S_j \text{ in } \{S_j\}_j\}.$$

The set $\mathcal{DP}_{\text{Prob}>0}(W)$ is given by

$$\mathcal{DP}_{\text{Prob}>0}(W) = \cup_j \mathcal{F}(P_j).$$

The value assignment to the members of $\mathcal{DP}_{\text{Prob}>0}(W)$ is given by

With probability $\text{Tr}(P_j W)$ the value of $Q \in \mathcal{F}(P_j)$ is
$[Q] = \text{Tr}(QP_j)/\text{Tr}(P_j)$,
and projections $\tilde{Q} \notin \mathcal{F}(P_j)$ do not have values.

The set $\mathcal{DP}_{\text{Prob}=1}(W)$ is given by

$$\mathcal{DP}_{\text{Prob}=1}(W) = \mathcal{F}(\{P_j\}_j).$$

The value assignment to the members of $\mathcal{DP}_{\text{Prob}=1}(W)$ is given by:

With probability $\text{Tr}(P_j W)$ the value of Q is $[Q] = \text{Tr}(QP_j)/\text{Tr}(P_j)$.

This value assignment to projections in $\mathcal{DP}_{\text{Prob}>0}(W)$ and $\mathcal{DP}_{\text{Prob}=1}(W)$ induces with a correspondence rule a value assignment to magnitudes pertaining to α and represented by hypermaximal operators A. If A has a spectral family $\{E_A(\Delta)\}_\Delta$ such that

$$A = \int_{\lambda=-\infty}^{\infty} \lambda dE_A((-\infty, \lambda]),$$

then the value $[A]$ of the magnitude represented by A is given by

$$[E_A(\Delta)] = 1 \leftrightarrow [A] \in^* \Delta$$
$$[E_A(\Delta)] = 0 \leftrightarrow [A] \notin^* \Delta.$$

The value assignment $[A] \in^* \Delta$ should be understood as "the value of A is restricted to the set Δ" and not as "$[A]$ has a precise value x and x is an element of Δ." The value assignment $[A] \notin^* \Delta$ should be understood as "the value of A is restricted to the set $\mathbb{R} - \Delta$." The properties of the value assignment to A are listed in section 10.

The new expanded property ascription assigns definite values to a larger set of projections than the property ascriptions proposed by Kochen (1985) and Clifton (1995a). Clifton assigns definite values to more projections than Kochen does (see section 4); the new property ascription assigns via $\mathcal{DP}_{\text{Prob}>0}(W)$ definite values to more projections than Clifton does.

The expanded property ascription endorses via $\mathscr{DP}_{\text{Prob}=1}(W)$ the conclusion of Clifton (1995a; 1995b) and Dickson (1995) that the set of projections that are definite-valued with probability 1 is equal to $\mathscr{F}(\{P_j\}_j)$. The expanded property ascription satisfies via $\mathscr{DP}_{\text{Prob}>0}(W)$ Healey's (1989) weakening condition and Arntzenius's (1990) intuition about measurement outcomes.

Appendix: The Conditional Property Ascription

In this appendix it is proved that the set $\mathscr{DP}(W,[P_a] = 1)$ of projections that have definite values according to the new expanded property ascription, given the state W and given $[P_a] = 1$, is equal to

$$\mathscr{DP}(W,[P_a] = 1) = \mathscr{F}(P_a) \tag{A.1}$$

with

$$[Q] = \text{Tr}(QP_a)/\text{Tr}(P_a). \tag{A.2}$$

A.1. Predictions with Lüders Measurements

First choose Lüders measurements for the calibrated measurements mentioned in the conditional certainty condition. In Vermaas (1996) probabilities are derived with which one can predict the outcomes of such Lüders measurements.

Take a magnitude pertaining to a system α and represented by an operator A^α defined in $\mathscr{H}^\alpha$. Assume that this operator has a spectral resolution $A^\alpha = \Sigma_{j,q}\, a_j |a^\alpha_{j,q}\rangle\langle a^\alpha_{j,q}|$ with distinct eigenvalues $\{a_j\}_j$ and mutually orthonormal eigenvectors $\{|a^\alpha_{j,q}\rangle\}_{j,q}$. A Lüders measurement of magnitude A^α is defined[19] as an interaction between α and a measurement device μ such that the evolution $U^{\alpha\mu}_{t_2,t_1}$ from t_1 to t_2 of the composite of α and μ obeys

$$\forall j,q\colon\quad U^{\alpha\mu}_{t_2,t_1}|a^\alpha_{j,q}\rangle|m^\mu_0\rangle = |a^\alpha_{j,q}\rangle|m^\mu_j\rangle.$$

Here, the projections $\{|m^\mu_k\rangle\langle m^\mu_k|\}_{k=0}$ are defined in the Hilbert space $\mathscr{H}^\mu$ associated with μ. The projection $|m^\mu_0\rangle\langle m^\mu_0|$ represents the initial ready-to-measure state of μ, and the orthogonal projections $\{|m^\mu_k\rangle\langle m^\mu_k|\}_{k=1}$ represent the pointer positions.

Consider a Lüders measurement of magnitude A^α. The initial state of α is some arbitrary $W^\alpha_{t_1}$, the initial state of μ is per definition $|m^\mu_0\rangle\langle m^\mu_0|$. On the basis of the core property ascription α possesses at t_1 the magnitudes $\{P^\alpha_j(t_1)\}_j$, and μ can possess at t_2 the outcomes $\{|m^\mu_k\rangle\langle m^\mu_k|\}_{k=1}$. The transition probability between an initial possessed magnitude $[P^\alpha_j(t_1)] = 1$ and a final outcome $[|m^\mu_k\rangle\langle m^\mu_k|] = 1$ is equal to[20]

$$\mathrm{Prob}([|m_k^\mu\rangle\langle m_k^\mu|] = 1/[P_j^\alpha(t_1)] = 1) = \frac{\mathrm{Tr}^\alpha(\Sigma_q |a_{k,q}^\alpha\rangle\langle a_{k,q}^\alpha| P_j^\alpha(t_1))}{\mathrm{Tr}^\alpha(P_j^\alpha(t_1))} \tag{A.3}$$

provided that the outcome $|m_k^\mu\rangle\langle m_k^\mu|$ is a possessed property at t_2. (The state $W_{t_2}^\mu$ is equal to $\Sigma_{c,d}\langle a_{c,d}^\alpha| W_{t_1}^\alpha |a_{c,d}^\alpha\rangle |m_c^\mu\rangle\langle m_c^\mu|$. So $|m_k^\mu\rangle\langle m_k^\mu|$ is an eigenprojection of $W_{t_2}^\mu$, but in order for $|m_k^\mu\rangle\langle m_k^\mu|$ to be a possessed magnitude, the corresponding eigenvalue $\Sigma_d \langle a_{k,d}^\alpha| W_{t_1}^\alpha |a_{k,d}^\alpha\rangle$ must be nondegenerate.)

A Lüders measurement of a projection Q^α is equivalent to measuring a magnitude A^α with eigenvalues $a_1 = 1$ and $a_2 = 0$ and eigenprojections $\Sigma_q |a_{1,q}^\alpha\rangle\langle a_{1,q}^\alpha| = Q^\alpha$ and $\Sigma_q |a_{2,q}^\alpha\rangle\langle a_{2,q}^\alpha| = \mathbb{1}^\alpha - Q^\alpha$. The outcome of this measurement of Q^α is either $[|m_1^\mu\rangle\langle m_1^\mu|] = 1$ or $[|m_2^\mu\rangle\langle m_2^\mu|] = 1$. The probabilities with which these outcomes are predicted can be determined with the transition probabilities (A.3). The first outcome is interpreted as Q^α having definite value 1, the second outcome is interpreted as Q^α having definite value 0. The probabilities with which one can predict the value of Q^α in a Lüders measurement are therefore

$$\begin{aligned} \mathrm{Prob}([Q^\alpha] = 1/[P_j^\alpha(t_1)] = 1) &= \frac{\mathrm{Tr}^\alpha(Q^\alpha P_j^\alpha(t_1))}{\mathrm{Tr}^\alpha(P_j^\alpha(t_1))} \\ \mathrm{Prob}([Q^\alpha] = 0/[P_j^\alpha(t_1)] = 1) &= \frac{\mathrm{Tr}^\alpha((\mathbb{1}^\alpha - Q^\alpha)P_j^\alpha(t_1))}{\mathrm{Tr}^\alpha(P_j^\alpha(t_1))} \end{aligned} \tag{A.4}$$

provided that $\mathrm{Tr}^\alpha(Q^\alpha W_{t_1}^\alpha) \neq \frac{1}{2}$. (If $\mathrm{Tr}^\alpha(Q^\alpha W_{t_1}^\alpha) = \frac{1}{2}$, μ doesn't possess pointer positions at t_2.)

A.2. Definite-valued Projections

Let's determine which projections get definite values according to the new expanded property ascription. Take a system α at t_1 with state $W_{t_1} = \Sigma_j w_j(t_1)P_j(t_1)$ and assume that $[P_a(t_1)] = 1$. First, on the basis of the core property ascription, the eigenprojections of the state W_{t_1} have definite values:

$$\{P_j(t_1)\}_j \subseteq \mathcal{DP}(W_{t_1}, [P_a(t_1)] = 1) \tag{A.5}$$

with

$$[P_a(t_1)] = 1 \quad \text{and} \quad [P_j(t_1)] = 0 \quad \text{if } j \neq a. \tag{A.6}$$

Second, on the basis of the conditional certainty condition, a number of projections Q have definite values. Suppose that Q is measured with a Lüders measurement from t_1 to t_2. If $\mathrm{Tr}(QW_{t_1}) = \frac{1}{2}$, the measurement device μ does not possess pointer positions, so one cannot determine whether Q has a definite value. But if $\mathrm{Tr}(QW_{t_1}) \neq \frac{1}{2}$, it follows with the formulae (A.4) that the outcome is with probability 1 equal to $[Q] = 1$ if and only if

$$\mathrm{Tr}(QP_a(t_1)) = \mathrm{Tr}(P_a(t_1)) \quad \Leftrightarrow \quad \mathrm{Tr}((\mathbb{1} - Q)P_a(t_1)) = 0.$$

It can be proved that this latter equality holds if and only if $QP_a(t_1) = P_a(t_1)$ (see the theorem at the end of this appendix). With the formulae (A.4) it also follows that the outcome is with probability 1 equal to $[Q] = 0$ if and only if

$$\mathrm{Tr}((\mathbb{1} - Q)P_a(t_1)) = \mathrm{Tr}(P_a(t_1)) \quad \Leftrightarrow \quad \mathrm{Tr}(QP_a(t_1)) = 0.$$

It can analogously be proved that this is the case if and only if $QP_a(t_1) = 0$. With the conditional certainty condition it thus follows that

$$\begin{aligned}\{Q \mid QP_a(t_1) = P_a(t_1) \ \text{ or } 0, \ \text{ and } \ \mathrm{Tr}(QW_{t_1}) \neq \tfrac{1}{2}\} \\ \subseteq \mathcal{DP}(W_{t_1}, [P_a(t_1)] = 1)\end{aligned} \tag{A.7}$$

with

$$\begin{aligned}[Q] = 1 \Leftrightarrow Q \in \{Q \mid QP_a(t_1) = P_a(t_1), \text{ and } \mathrm{Tr}(QW_{t_1}) \neq \tfrac{1}{2}\} \\ [Q] = 0 \Leftrightarrow Q \in \{Q \mid QP_a(t_1) = 0, \text{ and } \mathrm{Tr}(QW_{t_1}) \neq \tfrac{1}{2}\}.\end{aligned} \tag{A.8}$$

Third, on the basis of the closure condition, and the results stated above, one can prove (A.1) and (A.2). Explicit references to time are suppressed in the proof.

Let's start with proving (A.1). With the closure condition and the property ascriptions (A.5) and (A.7), it follows that $\mathcal{DP}(W,[P_a] = 1)$ is the closure under $\vee$, $\wedge$, and $\neg$ of the union of the sets $\{P_j\}_j$ and $\{Q \mid QP_a = P_a \text{ or } 0, \text{ and } \mathrm{Tr}(QW) \neq \tfrac{1}{2}\}$. First I prove that $\mathcal{DP}(W,[P_a] = 1)$ is a subset of $\mathcal{F}(P_a)$, and then I prove the converse.

With the definition (4.4) of a faux-Boolean algebra it is easily checked that the union of the sets $\{P_j\}_j$ and $\{Q \mid QP_a = P_a \text{ or } 0, \text{ and } \mathrm{Tr}(QW) \neq \tfrac{1}{2}\}$ is a subset of $\mathcal{F}(P_a)$. As a consequence the closure of this union is a subset of the closure of $\mathcal{F}(P_a)$. The closure of $\mathcal{F}(P_a)$ is $\mathcal{F}(P_a)$ itself because faux-Boolean algebras are already closed. It thus follows that $\mathcal{DP}(W,[P_a] = 1)$ is a subset of $\mathcal{F}(P_a)$.

To prove that $\mathcal{F}(P_a)$ is a subset of $\mathcal{DP}(W,[P_a] = 1)$, take any projection Q member of $\mathcal{F}(P_a)$. $\mathcal{F}(P_a)$ can be divided into the following three disjoint subsets:

$$\{Q \mid QP_a = P_a \text{ or } 0, \ \text{ and } \ \mathrm{Tr}(QW) \neq \tfrac{1}{2}\} \tag{A.9}$$

$$\{Q \mid QP_a = P_a, \ \text{ and } \ \mathrm{Tr}(QW) = \tfrac{1}{2}\} \tag{A.10}$$

$$\{Q \mid QP_a = 0, \ \text{ and } \ \mathrm{Tr}(QW) = \tfrac{1}{2}\}. \tag{A.11}$$

If Q is a member of the first subset (A.9), Q is also a member of $\mathcal{DP}(W,[P_a] = 1)$. If Q is a member of the second subset (A.10), Q can be written as

$$Q = Q(P_a + \mathbb{1} - P_a) = P_a + Q(\mathbb{1} - P_a).$$

The projections P_a and $Q(\mathbb{1} - P_a)$ are orthogonal so Q is equal to the join $P_a \vee \{Q(\mathbb{1} - P_a)\}$. The projection P_a is a member of $\mathcal{DP}(W,[P_a] = 1)$ due to (A.5). The projection $Q(\mathbb{1} - P_a)$ is a member of $\mathcal{DP}(W,[P_a] = 1)$ due to (A.7) because $\{Q(\mathbb{1} - P_a)\}P_a = 0$ and because $\mathrm{Tr}(\{Q(\mathbb{1} - P_a)\}W) = \frac{1}{2} - \mathrm{Tr}(P_aW) < \frac{1}{2}$ (this last inequality holds because $\mathrm{Tr}(P_aW) > 0$). Since P_a and $Q(\mathbb{1} - P_a)$ are members of $\mathcal{DP}(W,[P_a] = 1)$ and since $\mathcal{DP}(W,[P_a] = 1)$ is closed under the join $\vee$, Q is also a member of $\mathcal{DP}(W,[P_a] = 1)$.

If Q is a member of the third subset (A.11), Q can be written as $\neg\neg Q = \neg(\mathbb{1} - Q)$. The projection $(\mathbb{1} - Q)$ is a member of subset (A.10) and thus a member of $\mathcal{DP}(W,[P_a] = 1)$ and since $\mathcal{DP}(W,[P_a] = 1)$ is closed under the orthocomplementation $\neg$, Q is also a member of $\mathcal{DP}(W,[P_a] = 1)$.

So, any Q member of $\mathcal{F}(P_a)$ is also a member of $\mathcal{DP}(W,[P_a] = 1)$ and $\mathcal{F}(P_a)$ is thus a subset of $\mathcal{DP}(W,[P_a] = 1)$. □

The value assignment (A.2) can be proved as follows: Take any projection that is a member of $\mathcal{DP}(W,[P_a] = 1) = \mathcal{F}(P_a)$ and consider again the three subsets (A.9), (A,10), and (A.11).

If Q is a member of the first subset (A.9), the value of Q is given by formula (A.8). This value assignment is consistent with (A.2). If Q is a member of the second subset (A.10), Q is equal to $P_a \vee \{Q(\mathbb{1} - P_a)\}$. The value of P_a is with formula (A.6) equal to 1. The value of $Q(\mathbb{1} - P_a)$ is with formula (A.8) equal to 0. As a consequence the value of Q is equal to 1 and this value assignment is consistent with (A.2). If Q is a member of the third subset (A.11), Q is equal to $\neg(\mathbb{1} - Q)$ where the projection $(\mathbb{1} - Q)$ is a member of subset (A.10). Because $(\mathbb{1} - Q)$ is a member of subset (A.10), its value is 1. As a consequence the value of Q is equal to 0. This value assignment is also consistent with (A.2).

So, for any Q member of $\mathcal{DP}(W,[P_a] = 1)$ it follows that $[Q]$ is given by (A.2). □

A.3. A Theorem

The proof that $\mathrm{Tr}((\mathbb{1} - Q)P_a) = 0$ if and only if $QP_a = P_a$ is facilitated with the following.

Lemma

For each self-adjoint and positive operator P and each orthonormal set $\{|e_j\rangle\}_j$ of vectors, the following holds:

$$\langle e_j|P|e_j\rangle = 0 \quad \Rightarrow \quad \forall k: \qquad \langle e_j|P|e_k\rangle = \langle e_k|P|e_j\rangle = 0.$$

Proof

Because P is a positive operator it holds that $\langle\psi|P|\psi\rangle \geq 0$ for every vector $|\psi\rangle$. Take now $\langle e_j|P|e_j\rangle = 0$ and assume that $\langle e_j|P|e_k\rangle \neq 0$. Define the vector $|\psi\rangle = \lambda|e_j\rangle + |e_k\rangle$. Then

$$\langle\psi|P|\psi\rangle = \langle e_k|P|e_k\rangle + 2\ \mathrm{Re}[\bar{\lambda}\langle e_j|P|e_k\rangle].$$

If $\langle e_k|P|e_k\rangle = 0$, choose λ as

$$\lambda = -\frac{1}{\langle e_k|P|e_j\rangle}.$$

If $\langle e_k|P|e_k\rangle \neq 0$ (P is positive, so $\langle e_k|P|e_k\rangle \geq 0$), choose λ as

$$\lambda = -\frac{\langle e_k|P|e_k\rangle}{\langle e_k|P|e_j\rangle}.$$

In both cases $|\psi\rangle$ is such that $\langle\psi|P|\psi\rangle$ is negative, contradicting that P is positive. By *reductio ad absurdum* it follows that if $\langle e_j|P|e_j\rangle = 0$, then $\langle e_j|P|e_k\rangle = 0$. The complex conjugate of this result yields $\langle e_k|P|e_j\rangle = 0$. □

Theorem

$$\mathrm{Tr}((\mathbb{1} - Q)P_a) = 0 \quad \Leftrightarrow \quad QP_a = P_a$$

Proof

The $\Leftarrow$ part of the proof is trivial so I only consider the $\Rightarrow$ part.

Let $\{|q_{0,j}\rangle\}_j$ be a complete set of orthonormal eigenvectors of Q with eigenvalue 0. Let $\{|q_{1,k}\rangle\}_k$ be a complete set of orthonormal eigenvectors of Q with eigenvalue 1. Q has only eigenvalues 0 and 1 so the union of these two sets of eigenvectors is a complete orthonormal basis of $\mathcal{H}^\alpha$.

Perform now the trace over $(\mathbb{1} - Q)P_a$ by means of this complete orthonormal basis $\{|q_{0,j}\rangle\}_j \cup \{|q_{1,k}\rangle\}_k$. It follows that

$$\mathrm{Tr}((\mathbb{1} - Q)P_a) = 0 \Leftrightarrow \sum_j \langle q_{0,j}|P_a|q_{0,j}\rangle = 0.$$

P is a positive operator so each term in this sum is either 0 or larger than 0, hence

$$\forall j: \langle q_{0,j}|P_a|q_{0,j}\rangle = 0.$$

One can now employ the lemma and conclude that

$$\begin{aligned} \forall j,j'\colon \langle q_{0,j}|P_a|q_{0,j'}\rangle &= 0 \\ \forall j,k\colon \langle q_{0,j}|P_a|q_{1,k}\rangle &= 0. \end{aligned} \tag{A.12}$$

With these identities it can be shown that QP_a and P_a are equal because the matrix elements of these operators with respect to the basis $\{|q_{0,j}\rangle\}_j \cup \{|q_{1,k}\rangle\}_k$ are equal:

$$\begin{aligned} \forall j,j'\colon \langle q_{0,j}|QP_a|q_{0,j'}\rangle &= \langle q_{0,j}|P_a|q_{0,j'}\rangle \\ \forall j,k\colon \langle q_{0,j}|QP_a|q_{1,k}\rangle &= \langle q_{0,j}|P_a|q_{1,k}\rangle \\ \forall j,k\colon \langle q_{1,k}|QP_a|q_{0,j}\rangle &= \langle q_{1,k}|P_a|q_{0,j}\rangle \\ \forall k,k'\colon \langle q_{1,k}|QP_a|q_{1,k'}\rangle &= \langle q_{1,k}|P_a|q_{1,k'}\rangle. \end{aligned}$$

The first two equations hold because the left-hand sides are equal to 0 with $\langle q_{0,j}|Q = 0$ and the right-hand sides are equal to 0 due to (A.12). The last two equations hold because $\langle q_{1,k}|Q = \langle q_{1,k}|$. □

Notes

I am indebted to Dennis Dieks for valuable comments on earlier versions of this chapter. Also I would like to thank Rob Clifton and Jos Uffink for discussions, and the editors for their invitation to contribute to this volume. This work was supported by the Netherlands Organization for Scientific Research (NWO).

1. This version of the modal interpretation is a generalization of the version introduced by Kochen (1985), Dieks (1988), and Healey (1989).

2. In this spectral resolution the sum runs over all different *nonzero* eigenvalues w_j of W and the P_j's are eigenprojections of W. An eigenprojection P_j is multidimensional if the eigenvalue w_j is degenerate. The spectral resolution is unique in the sense that the sets $\{w_j\}_j$ and $\{P_j\}_j$ are uniquely determined by W.

3. See section 4 of Kochen (1985).

4. I actually simplified history a bit. Kochen (1985) considered the property ascription to α only if the spectral resolution of the state W is nondegenerate. The eigenprojections $\{P_j\}_j$ of W are then one-dimensional and Kochen assigned values to them. In the case that the spectral resolution of W is degenerate, Kochen left the value assignment unspecified. Healey (1989) and Dieks (1993) extended the modal interpretation in order to cover the latter case. If W has a degenerate eigenvalue w_k with a corresponding multidimensional eigenprojection P_k, a value is assigned to this multidimensional projection. In both cases the value assignment is captured by the core property ascription.

5. The lattice operations are defined as usual. Let P and Q be projections onto the subspaces $\mathcal{P}$ and $\mathcal{Q}$ of $\mathcal{H}^\alpha$, respectively. The join $P \vee Q$ is then the projection onto the subspace $\mathcal{P} \otimes \mathcal{Q}$, the meet $P \wedge Q$ is the projection onto $\mathcal{P} \cap \mathcal{Q}$, and the orthocomplement $\neg P$ the projection onto the subspace $\perp \mathcal{P}$, which is the projection $\mathbb{1} - P$.

6. See section 3 of Clifton (1995a).

7. The original definition of a faux-Boolean algebra in Dickson (1995), section 1, is slightly different from the one presented here. Dickson defines a faux-Boolean algebra as the closure of the union of $\{S_j\}_j$ and the set of all *one*-dimensional projections $\{Q\}$ orthogonal to all S_j. Both definitions are clearly equivalent.

8. See section 4 of Clifton (1995a).

9. See page 39 of Clifton (1995a).

10. By a properly calibrated measurement I mean a measurement that reproduces the predictions of quantum theory. If a measurement of a projection Q is performed on a system α with an initial state W, then it is required that the expectation value $\mathrm{Tr}^{\mu}(M_1^{\mu} W^{\mu})$ is equal to $\mathrm{Tr}(QW)$. Here W^{μ} is the state of the measurement device μ after the measurement and M_1^{μ} is the pointer position of μ that corresponds to Q. For an elaborated description of calibrated measurements, see Busch, Lahti, and Mittelstaedt (1991), especially the survey and section 2.3 of chapter 3.

11. The certainty condition should be distinguished from the condition Einstein, Podolsky, and Rosen (1935) introduced for the existence of an element-of-reality. With both conditions one can infer from the result of a measurement a definite value of a magnitude pertaining to a system α. However, the certainty condition requires that the measurement is performed on α itself and allows a disturbance of the state of α during the measurement. The Einstein, Podolsky, and Rosen condition, in its turn, requires that the state of α not be disturbed but allows that the measurement be performed on a system different from α.

12. See section 2.2 of Healey (1989).

13. See section 3, problem 1, of Arntzenius (1990).

14. Lüders measurements are indeed calibrated measurements, for they are a special type of normal unitary premeasurements; see chapter 3 of Busch, Lahti, and Mittelstaedt (1991). Lüders measurements are discussed more extensively in the appendix.

15. To be precise, the tacit assumption is not that for all $P \in \mathcal{DP}$ the probability that $[P] = 0$ is equal to 1, or that the probability that $[P] = 1$ is 1; the assumption is that the sum of these two probabilities is equal to 1.

16. See page 43 of Clifton (1995a).

17. $\mathcal{DP}_{\mathrm{Prob}>0}$ does not obey the closure condition. Take the state W given in equation (6.1). The projections

$$Q = |w_1\rangle\langle w_1| + \tfrac{1}{2}(|w_2\rangle + |w_3\rangle)(\langle w_2| + \langle w_3|)$$
$$\tilde{Q} = |w_2\rangle\langle w_2| + \tfrac{1}{2}(|w_1\rangle + |w_3\rangle)(\langle w_1| + \langle w_3|)$$

are both members of $\cup_{j=1}^{3}\mathcal{F}(|w_j\rangle\langle w_j|)$. But their meet

$$Q \wedge \tilde{Q} = \tfrac{1}{3}(|w_1\rangle + |w_2\rangle + |w_3\rangle)(\langle w_1| + \langle w_2| + \langle w_3|)$$

is not a member of $\cup_{j=1}^{3}\mathcal{F}(|w_j\rangle\langle w_j|)$.
The set $\mathcal{DP}_{\mathrm{Prob}>0}$ does obey the weaker condition that the sets of projections that can have simultaneously definite values are closed under $\vee, \wedge, \neg$. With the value assignment (8.4) to $\mathcal{DP}_{\mathrm{Prob}>0}$, these sets are equal to $\mathcal{F}(P_j)$ and faux-Boolean algebras are indeed closed under $\vee$, $\wedge$, and $\neg$.

18. I adopt the choice negation mainly because one then obtains the relation "$[\neg E_A(\Delta)] = 1$ if and only if not $[A] \in^* \Delta$," that is, the map from value assignments to members of $\{E_A(\Delta)\}$ to value assignments to sets $\{\Delta\}$ preserves the logical structure of $\{E_A(\Delta)\}$.

19. See section 2.3 of chapter 3 of Busch, Lahti, and Mittlestaedt (1991).

20. See section 6 of Vermaas (1996).

References

Albert, D., and B. Loewer. 1990. "Wanted Dead or Alive: Two Attempts to Solve Schrödinger's Paradox." In A. Fine, M. Forbes, and L. Wessels, eds., *Proc. 1990 Biennial*

Meeting of the Philosophy of Science Association 1. East Lansing: Philosophy of Science Association, 277–85.

Arntzenius, F. 1990. "Kochen's Interpretation of Quantum Mechanics." In A. Fine, M. Forbes, L. Wessels, eds., *Proc. 1990 Biennial Meeting of the Philosophy of Science Association* 1. East Lansing: Philosophy of Science Association, 241–49.

Bacciagaluppi, G., and M. Hemmo. 1996. "Modal Interpretations, Decoherence, and Measurements." *Studies in History and Philosophy of Modern Physics* 27: 239–77.

Busch, P., P. J. Lahti, and P. Mittelstaedt. 1991. *The Quantum Theory of Measurement*. Berlin: Springer.

Clifton, R. 1995a. "Independent Motivation of the Kochen-Dieks Modal Interpretation of Quantum Mechanics." *British Journal for the Philosophy of Science* 46: 33–57.

———. 1995b. "Making Sense of the Kochen-Dieks 'No-Collapse' Interpretation of Quantum Mechanics Independent of the Measurement Problem." In D. Greenberger and A. Zeilinger, eds., *Fundamental Problems in Quantum Theory*. Annals of the New York Academy of Science, 755: 570–78.

Dickson, W. M. 1995. "Faux-Boolean Algebras, Classical Probabilities, and Determinism." *Foundations of Physics Letters* 8: 231–42.

Dieks, D. 1988. "The Formalism of Quantum Theory: An Objective Description of Reality?" *Annalen der Physik* 7: 174–90.

———. "The Modal Interpretation of Quantum Mechanics and Some of Its Relativistic Aspects." *International Journal of Theoretical Physics* 32: 2363–75.

Einstein, A., B. Podolsky, and N. Rosen. 1935. "Can Quantum-Mechanical Description of Physical Reality Be Considered Complete?" *Physical Review* 47: 777–80.

Healey, R. 1989. *The Philosophy of Quantum Mechanics*. Cambridge: Cambridge University Press.

Kochen, S. 1985. "A New Interpretation of Quantum Mechanics." In P. Lahti and P. Mittelstaedt, eds., *Symposium on the Foundations of Modern Physics*. Singapore: World Scientific, 151–69.

Kochen, S., and E. Specker. 1967. "The Problem of Hidden Variables in Quantum Mechanics." *Journal of Mathematics and Mechanics* 17: 59–87.

Vermaas, P. E. 1996. "Unique Transition Probabilities in the Modal Interpretation." *Studies in History and Philosophy of Modern Physics* 27: 133–59.

Vermaas, P. E., and D. Dieks. 1995. "The Modal Interpretation of Quantum Mechanics and Its Generalization to Density Operators." *Foundations of Physics* 25: 145–58.

Dennis Dieks

Preferred Factorizations and Consistent Property Attribution

1. Introduction

The two leading ideas of the version of the modal interpretation that is our starting point here are

1. Quantum mechanics should be interpreted in terms of *properties* possessed by physical systems, so that the "measurement problem" no longer occurs (measurement results are treated as properties possessed by a measuring device);

2. The properties assigned to a system should be definable from the quantum-mechanical state—no additional structure should be introduced (in this sense the interpretation is not a "hidden-variables interpretation"). (See Dieks 1994, 1995; Vermaas and Dieks 1995.)

In order to implement the idea that physical properties should be attributed, without getting into "Kochen and Specker difficulties," a *subset* of all observables in the Hilbert space associated with a physical system are selected as representing physical magnitudes possessing a definite value. Specifically, let α be our system and let β represent its total environment (the rest of the universe). Let $\alpha\&\beta$ be represented by $|\psi^{\alpha\beta}\rangle \in \mathcal{H}^{\alpha} \otimes \mathcal{H}^{\beta}$. The biorthonormal decomposition of $|\psi^{\alpha\beta}\rangle$,

$$|\psi^{\alpha\beta}\rangle = \sum_i c_i \, |\psi_i^{\alpha}\rangle \, |\psi_i^{\beta}\rangle, \tag{1.1}$$

with $\langle\psi_i^{\alpha}|\psi_j^{\alpha}\rangle = \langle\psi_i^{\beta}|\psi_j^{\beta}\rangle = \delta_{ij}$, generates two sets of projectors operating on $\mathcal{H}^{\alpha}$ and on $\mathcal{H}^{\beta}$, respectively: $\{|\psi_i^{\alpha}\rangle\langle\psi_i^{\alpha}|\}_i$ and $\{|\psi_i^{\beta}\rangle\langle\psi_i^{\beta}|\}_i$. If there is no degeneracy among the numbers $\{|c_i|^2\}$, these sets of one-dimensional projectors are uniquely determined by the decomposition. In the case of degeneracy the projectors belonging to one value of $\{|c_i|^2\}$ can be added to form a multidimensional projector; the new set of projectors, including multidimensional ones, is again uniquely determined. The projectors in question are the ones occurring in the spectral decomposition of the reduced density operators of α and β.

The modal interpretation assigns definite values to the physical magnitudes represented by these projectors, and to all functions of them ("well-defined" or "applicable" physical magnitudes, in Bohrian parlance). *Which* value among the possible values of a definite magnitude is actually realized is not fixed by the interpretation. For each possible value a probability is specified: the probability that the magnitude represented by $|\psi_i^\alpha\rangle\langle\psi_i^\alpha|$ has the value 1 is given by $|c_i|^2$. In the case of degeneracy it is stipulated that the magnitude represented by $\Sigma_{i\in I_l}|\psi_i^\alpha\rangle\langle\psi_i^\alpha|$ has value 1 with probability $\Sigma_{i\in I_l}|c_i|^2$ (I_l is the index-set containing indices j, k such that $|c_j|^2 = |c_k|^2$).

The above prescription for defining sets of definite magnitudes presupposes a factorization of the total Hilbert space. If a given total Hilbert space can be factorized, however, it can be factorized in infinitely many ways. The assumption that each of these factorizations corresponds to a subdivision of the total system into physical subsystems to which physical properties can be assigned in the above way, leads to unattractive consequences. As Bacciagaluppi has pointed out (1995), the same vector in $\mathcal{H}^\alpha \otimes \mathcal{H}^\beta$ can be a product vector in two different factorizations: $|\psi^\alpha\rangle|\psi^\beta\rangle = |\psi^\gamma\rangle|\psi^\delta\rangle$. But, as he has shown by means of a variation on the Kochen and Specker argument, it cannot always be true that if $|\psi^\alpha\rangle\langle\psi^\alpha|$ and $|\psi^\beta\rangle\langle\psi^\beta|$ have the value 1, the same holds for $|\psi^\gamma\rangle\langle\psi^\gamma|$ and $|\psi^\delta\rangle\langle\psi^\delta|$. This result, which in itself already appears paradoxical, leads to concrete difficulties when ideal measurements of $|\psi^\alpha\rangle\langle\psi^\alpha| \otimes |\psi^\beta\rangle\langle\psi^\beta| = |\psi^\gamma\rangle\langle\psi^\gamma| \otimes |\psi^\delta\rangle\langle\psi^\delta|$ are considered: it is impossible for the ideal measurement to reveal preexisting values both in $\alpha\&\beta$ and in $\gamma\&\delta$ (see Bacciagaluppi [1995] for more details about this "Bacciagaluppi-Kochen-Specker" paradox).

One way to avoid these awkward consequences is to assume that there is a preferred factorization of the total Hilbert space in terms of factor spaces which are associated with elementary physical systems. The idea is that most of the infinitely many other factorizations do not correspond to subdivisions of the total system into physical systems, but represent unphysical "mixtures" of the atomic physical systems. If this is accepted, it follows that arbitrary rotations of a product basis in a physical factorization of the Hilbert space no longer generate basis vectors of a product basis of some other physical factorization; this blocks the construction of the Kochen-Specker paradox (Bacciagaluppi 1995).

Of course, the assumption of basic atomic constituents fits in with standard physical conceptions about the nature of matter. Nevertheless, it is not possible to exclude other factorizations than a given physical one on the basis of the mathematical structure of Hilbert space alone; something more is needed. A natural further ingredient to consider is the Hamiltonian and other representatives of the space-time behavior of the system. We shall indeed propose to use requirements about the representation of the space-time group to exclude factorizations rival to an initially specified physical one.

The next purpose of this chapter is to show that if the existence of a preferred factorization is accepted, the above-mentioned prescription for ascribing properties and probabilities can be generalized in a very simple way to yield a joint probability distribution for the properties of all possible subsystems of a composite system. The recipe is simply to assign properties to all atomic systems and to postulate that the properties of non-atomic systems follow by "property composition" (Clifton [1995]). This way of assigning properties leads to a joint probability distribution with the right marginal distributions, and is obviously consistent. But it deviates from earlier proposals in that it does not allow for the possibility of "emergent" properties; that is, properties of composite systems that are not reducible to properties of their parts. At first sight this might seem a fatal shortcoming: it has become more or less common wisdom (at least among "modal theorists") that it is essential for quantum mechanics that systems can possess properties that do not supervene on the properties of their parts. We will argue, however, that the situation is not that clear. The simple interpretational scheme we just mentioned may very well be viable.

2. The Atomic Hilbert Spaces

We take our starting point in the situation that one factorization, carrying irreducible representations of the space-time transformation group, is given; with respect to this factorization a Hamiltonian H has been specified. We are also going to suppose that the dynamical theory is unique, in the sense that it is not possible to change the form of the Hamiltonian and still have a viable dynamical scheme. This is a strong supposition (perhaps a stronger one than is actually needed); it doubtless needs justification. We cannot provide that justification here, and only remark that the present situation in theoretical physics appears to show that consistency requirements restrict the possible types of physical interactions very severely. However that may be, our program here is to make it plausible that requirements on the space-time behavior of physical systems single out the given factorization as the only physical one; and we will use rather strong requirements to make the idea clear.

In nonrelativistic quantum mechanics H has the following form:

$$H = \sum_i p_i^2/2m_i + \sum_{i \neq j} H_{ij}, \tag{2.1}$$

where p_i is the momentum operator belonging to subsystem i, m_i the mass of this subsystem, and H_{ij} the interaction Hamiltonian of subsystems i and j. We are interested in possible ways to split up the total physical universe; we therefore do not include external potentials in the Hamiltonian.

Our initial assumption thus is that the Hamiltonian comes together with a physical factorization of the total Hilbert space; the factor spaces are the Hilbert spaces of the subsystems i, j, and so on. There is a representation of the Galilei group defined on each of the factor spaces and the dynamical variables (momentum, position, energy, and the like) have a double role, in that they also generate space-time transformations: p_i is the generator of translations, $m_i r_i$ the generator of Galilean boosts, and so on. The question now is whether it is possible to split up the system in other ways; in other words, are there physically significant *rival* factorizations of the total Hilbert space?

From a mathematical point of view there are of course other factorizations; some of them certainly possess physical relevance. An obvious example for the case of a two-particle system is the factorization associated with the center of mass and the "relative particle." If $H = p_1^2/2m_1 + p_2^2/2m_2 + V(r_1 - r_2)$, the change of variables

$$M = m_1 + m_2 \qquad R = \frac{m_1 r_1 + m_2 r_2}{m_1 + m_2} \qquad P = p_1 + p_2 \tag{2.2}$$

$$m = \frac{m_1 m_2}{m_1 + m_2} \qquad r = r_1 - r_2 \qquad p = \frac{m_2 p_1 - m_1 p_2}{m_1 + m_2} \tag{2.3}$$

transforms the Hamiltonian into $H = P^2/2M + p^2/2m + V(r)$. This transformation is canonical, so P, R and p, r have the characteristic commutation relations of particle momentum and position.

Nevertheless, it is common in the physics literature to refer to the two particles introduced in this way as "fictitious particles." The intuition is clearly that the "center of mass particle" and its "relative" counterpart can be introduced for the sake of mathematical convenience, but are not physically real in the same way as the original particles. But this intuition seems never to have been elaborated on so as to provide a formal criterion for distinguishing real systems from fictitious ones.

Two related points stand out as making the above fictitious particles different from the original ones. The Hamiltonian does not depend on the "mutual particle distance" $R - r$, so it is not translation invariant in the new coordinates, and the original total momentum $p_1 + p_2$ does not equal the new total momentum $P + p$. These points are related because momentum is the infinitesimal generator of translations. If we translate the total two-particle system over a distance a, the two original particles are both displaced over the same distance; the total translation is generated by $p_1 + p_2$ and the individual translations by p_1 and p_2, respectively. But it can be seen from equation (2.3) that the "relative particle" is not shifted at all; only the "center of mass particle" is subject to a displacement. This means that P and p do not generate physical displacements.

This suggests the following necessary condition for factorizations to correspond to physically real systems. The factor Hilbert spaces should carry a representation of the space-time group (the Galilei group in our example of nonrelativistic quantum mechanics) in the same way as the factor spaces of the original factorization, with the usual identification of generators of the space-time group and dynamical variables. This proposal fits in with the standard approach in elementary particle physics, according to which elementary particles are associated with irreducible representations of the space-time group.

The criterion is intended to imply, among other things, that the total generators of space-time transformations should be the sum of the corresponding generators in the individual factor spaces. For example, for physically significant factorizations the total generator of translations should be $\Sigma_i p_i$, and the generator of boosts should be $\Sigma_i G_i$, where the p_i (momenta) are the generators of translations for the individual systems; the G_i are the generators of Galilei boosts, $G_i = m_i r_i$. This expresses mathematically the idea that all physical systems are "embedded" in space-time in the same way. In a spatial translation of the total system over a certain distance all individual partial systems should be translated over that same distance (compare the above example of the two fictitious particles) and in an overall change of velocity (as effected in a transition to a moving frame of reference) all individual systems should undergo the same velocity change; and similarly for rotations. It seems that this embedding in space-time is a central aspect of what we understand by real physical systems (as opposed to fictitious ones). The conditions $G = \Sigma_i G_i$ and $P = \Sigma_i p_i$, and similar ones for other generators, guarantee that all parts of the system are affected in the same way by global space-time transformations. For example, if $P = \Sigma_i p_i$, the unitary operator which represents a translation over a distance a has the form $U(a) = \Pi_i U_i(a)$, where $U_i(a)$ is a unitary operator representing a translation over a in factor space i.

We will therefore look for factorizations with respect to which there are generators $\{p'_j\}$, $\{G'_j\}$, etc., such that $\Sigma_i p_i = \Sigma_j p'_j$, $\Sigma_i G_i = \Sigma_j G'_j$, and with respect to which the form of the Hamiltonian is the same as in the original factorization. (Another possible requirement would simply be that the dynamical variables always also are space-time group generators; perhaps this would be sufficient. We take the rather strong invariance requirement on the Hamiltonian to make the following simple semiclassical discussion possible.) It is simplest and sufficient for our purpose here to consider the classical analogues of our systems; if canonical changes of the classical variables exist that satisfy our requirements, the standard quantization procedure will lead to alternative factorizations in the quantum case.

First consider the case of two free particles: $H = p_1^2/2m_1 + p_2^2/2m_2$. Take two positive numbers m_1' and m_2', such that $m_1' + m_2' = m_1 + m_2 = M$. (The total mass of the system is fixed by its global space-time behavior.) In order that new coordinates r_1' and r_2' have the required transformation properties under rotations and translations, we should have

$$r_1' = a_1 r_1 + a_2 r_2 \qquad a_1 + a_2 = 1 \tag{2.4}$$

$$r_2' = b_1 r_1 + b_2 r_2 \qquad b_1 + b_2 = 1. \tag{2.5}$$

It can easily be checked that the following values of a_1, b_1, a_2, and b_2 leave the form of H invariant.

$$a_1 = \frac{m_1}{M} + \frac{\sqrt{m_1 m_2 m_1' m_2'}}{M m_1'} \qquad b_1 = \frac{m_1}{M} - \frac{\sqrt{m_1 m_2 m_1' m_2'}}{M m_2'} \tag{2.6}$$

$$a_2 = \frac{m_2}{M} - \frac{\sqrt{m_1 m_2 m_1' m_2'}}{M m_1'} \qquad b_2 = \frac{m_2}{M} - \frac{\sqrt{m_1 m_2 m_1' m_2'}}{M m_2'}. \tag{2.7}$$

With this choice of the coefficients we have

$$H = p_1^2/2m_1 + p_2^2/2m_2 = p_1'^2/2m_1' + p_2'^2/2m_2'. \tag{2.8}$$

A system of free particles can therefore be split up in indefinitely many ways while keeping the Hamiltonian and the space-time properties the same; the total mass M can be divided arbitrarily in parts m_1' and m_2'. For the quantum case this means that there are infinitely many factorizations, each representing free particles, but with different masses. There is no unique subdivision of the system.

The interesting (because realistic) case is that in which there is interaction between the partial systems; that is, in which $H_{12} = V(r_2 - r_1) \neq 0$. The requirement that the interaction term in the Hamiltonian keep the same form in a transition to the new coordinates now leads to an additional constraint on the new variables. We have for the transformation of $r_2 - r_1$

$$r_2' - r_1' = (a_1 - b_1)(r_2 - r_1) = \sqrt{\frac{m_1 m_2}{m_1' m_2'}}(r_2 - r_1). \tag{2.9}$$

Because of its universal character, the gravitational interaction is an efficient test case: $V = Gm_1m_2/r$. This will only be invariant under the above transformation if $m_1m_2 = m_1'm_2'$. In view of the fact that also $m_1 + m_2 = m_1' + m_2'$, this immediately leads to $m_1 = m_1'$, $m_2 = m_2'$, and $a_1 = 1$, $a_2 = 0$, $b_1 = 0$, $b_2 = 1$ (and the uninteresting other solution that consists of an exchange of m_1 and m_2).

This means that it is not possible to subdivide the total system in other ways (given the standard division) such that the given standard form of dynamical theory is maintained (including the standard value of constants like G).[1] Given one standard factorization of the total Hilbert space, and a dynamical theory (involving interactions) defined with respect to that factorization, there is no other factorization with the same space-time properties.

3. Joint Property Attribution

We consider a composite system represented in the Hilbert space $\mathcal{H}^\alpha \otimes \mathcal{H}^\beta \otimes \ldots \otimes \mathcal{H}^\theta \otimes \ldots \otimes \mathcal{H}^\xi$, where the atomic factor spaces $\mathcal{H}^\alpha$, $\mathcal{H}^\beta, \ldots, \mathcal{H}^\xi$ are associated with the atomic subsystems denoted by α, $\beta, \ldots \xi$. Of course, there are also bigger subsystems, represented in tensor products of a number of atomic spaces. These systems we denote by $\alpha\&\beta$, $\gamma\&\delta\&\theta$, and so on. We assume that the total system is represented by a pure state. This assumption facilitates the following discussion, but our results will also cover the case in which the total state is a mixture.

According to the rule for ascribing properties discussed in section 1, a set of definite-valued projectors is assigned to each atomic subsystem: $\{P_i^\alpha\}_i$, $\{P_j^\beta\}_j$, and so on. The rule as given in section 1 only specifies the joint probabilities for properties of two systems. But it is easy to formulate a generalization which applies to the case of an arbitrary number of atomic systems (Vermaas and Dieks 1995):

$$\begin{aligned}\text{Prob}(P_i^\alpha, P_j^\beta, \ldots P_k^\theta, \ldots P_l^\xi) \\ = \langle \Psi | P_i^\alpha \otimes P_j^\beta \otimes \ldots P_k^\theta \otimes \ldots P_l^\xi | \Psi \rangle,\end{aligned} \tag{3.1}$$

where the left-hand side represents the probability for the various projectors to take the value 1. It is crucial for the consistency of this probability ascription that the projection operators occurring in the formula all commute (which they of course do, since they operate in different atomic Hilbert spaces).

According to most versions of the modal interpretation proposed up to now, an important characteristic of the property attribution is that the properties of composite systems, such as $\alpha\&\beta$, can be "emergent"; that is, they do not always supervene on the properties of their parts (Dieks 1989, 1994; Healey 1989) (however, Bacciagaluppi and Dickson [1995] contains a suggestion along the lines of what we will discuss here). This means that it is in general not true that $P_i^{\alpha\&\beta} = P_k^\alpha \otimes P_l^\beta$, for some value of k and l. That not all properties of nonatomic systems are "built up" from properties of atomic systems is a consequence of applying the biorthogonal decomposition rule not only to atomic systems, but also to composite ones. For example, the total system, represented by $|\Psi\rangle$, thus is assigned the property corresponding to $|\Psi\rangle\langle\Psi|$, whereas $|\Psi\rangle\langle\Psi| \neq P_i^\alpha \otimes P_j^\beta \otimes \ldots \otimes P_k^\xi$. An important consequence of assigning properties this way (that is, by applying the biorthogonal decomposition rule to *all* subsystems) is that the definite-valued projectors of overlapping systems (subsystems of the total system) will in general not commute. This makes it impossible to generalize equation (3.1) by simply inserting in the product the projectors pertaining to the composite systems; the resulting operator, of which the expectation value is taken, would no longer be hermitean, and the "probabilities" would not always be positive real numbers. It would therefore be necessary to find some other joint probability distribution for composite systems with "emergent" properties. No natural candidate expression offers itself, however. As long as this problem has not been solved, the consistency of the joint property attribution is not beyond doubt. Can we be sure that a joint distribution exists which possesses all the right marginal distributions?

In order to circumvent this problem we now make a radical suggestion. We propose to apply the biorthogonal decomposition rule to atomic systems only; all larger systems inherit the properties of their atomic constituents by property composition. That is, we assume that the definite-valued projectors of $\alpha\&\beta$ are given by $P_i^\alpha \otimes P_j^\beta$, with P_i^α and P_j^β definite-valued projectors of α and β, respectively (except, perhaps, in the case where the atomic constituents only have the trivial property represented by the projector on their entire Hilbert space—a point to be commented on later).

This proposal closely follows classical ideas about physical ontology. In classical physics all properties of systems supervene on the properties of their elementary constituents (atoms, molecules). But it has become a widely accepted notion that quantum mechanics requires a certain "holism," according to which systems can have properties that are *not* reducible to properties of their parts. There are many examples in which this feature appears to be essential. These cases are characterized by the circumstance that the state of the larger system is a superposition of eigenstates belonging to the definite projectors of the composing atomic systems.

4. "Emergent Properties"

The reason for assuming the existence of "emergent properties" is that quantum mechanics often attributes states to composite systems that are not product states. These states play a very significant role in securing the empirical adequacy of the theory. An example is furnished by the quantum-mechanical treatment of the ammonia molecule (Feynman, Leighton, and Sands 1965). This molecule consists of three hydrogen atoms and one nitrogen atom (consisting of other components in their turn), in a tetradic structure. But the molecule can also be treated on a more global level. In important situations the total state is a superposition, with time-dependent coefficients, of two states (corresponding to tetradic structures which are each other's mirror images relative to the plane of the hydrogen atoms). These two states are associated with opposite electric dipole moments. In an electric field the molecule therefore essentially behaves as a two-state system (comparable to a single spin$-\frac{1}{2}$ particle). This makes it possible to explain the working of the ammonia maser: the molecule exchanges energy quanta with the field, corresponding to the energy difference between the two states.

The just-mentioned description, which is the one used in explaining the radiation emitted by the molecule, is on the level of the molecule as a whole. Suppose, however, that the Hamiltonian we use contains the free Hamiltonians of the atoms plus their interactions with each other and the electric field. If we adopt the proposals of sections 2 and 3, we then have to describe the system as composite, built up from hydrogen and nitrogen atoms in interaction with the field modes. According to this description the ammonia molecule is not an independent entity but inherits its properties from its atomic constituents; this means that we cannot say that the molecule has either the "spin-up" or "spin-down" property (which we *could* say in a simple model in which the molecule is treated as one atomic system). However, *for the properties ascribed to the field modes this makes no difference*. These properties are determined by the biorthogonal decomposition that relates to the division: field mode versus rest of the system. In other words, the properties assigned to the field do not reflect the individual properties of the constituents of the molecule, but rather represent the result of the *total* interaction with the molecule. This implies that the working of the maser is still adequately covered, in the treatment in which the ammonia molecule is not an atomic system. More generally, all information about the ammonia system that is conveyed through the field remains the same as in the case in which the molecule is treated as one whole.

The following picture seems appropriate. The electric field interacts with the molecule, but because of collective effects the result is not just the addition of several individual interactions. Consequently, the state of the field does not reflect the properties of the individual constituents of the molecule. There is a *collective dynamical effect* on the field. The quantitative details of this effect follow from the properties assigned to the field by means of the decomposition of the total quantum state in terms of the division field versus rest of the world.

Central in this account is the denial that the "relative state" associated with a property of the field corresponds to a property of the molecule as a whole. The molecule is considered to be an assembly of atoms, and there are no properties of the molecule over and above the properties of the members of the assembly. But the molecule can have typically quantum-mechanical collective effects on the outside world, which can be calculated from the dynamics and the total quantum-mechanical state.

The surprising element in this proposal is that collective effects need not correspond to collective properties. Still, to some extent there is an analogy here with classical physics. Also in classical physics a description in terms of composing particles rules out that a description that uses "collective properties" has an equally fundamental status. Consider the classical situation in which a force affects a system "through its center of mass." If the system is treated as being built up from more elementary particles, the reference to the center of mass as the point in space where the force operates is only a *façon de parler* that does not involve the assumption that the center of mass is itself the position of a physical system. Indeed, from the point of view according to which the system is composite, the center of mass will usually find itself in empty space; if we consider the system to consist of particles there usually *is* nothing at the position of the center of mass. Speaking about the center of mass, and "the total mass which is located there," is justified by an associated simple way of treating the *collective effect* of the particles on their surroundings.

If we don't know about the internal constitution of the body (or are not interested in it), we may replace the fine-grained description by one in which only the center of mass is used. Within that coarse-grained description, which employs a less detailed and simpler Hamiltonian, the center of mass system assumes the status of a real physical system. We might express this familiar state of affairs in a rather spectacular way by saying that different attributions of physical reality arise, relative to the level of detail contained in the Hamiltonian. The coarse-grained and fine-grained descriptions are mutually exclusive in their attributions of physical reality. What is a *property* of the system according to the coarse-grained description, only has the significance of something playing a role in the calculation of a *collective effect* in the fine-grained description.

A similar "complementarity of levels of description" (though things are naturally more fundamental here than in the classical analogue) arises in the present proposal of an "atomic version" of the modal interpretation. If a Hamiltonian for the ammonia molecule is given that only contains the electric dipole interaction with the field, with the consequence that the internal structure of the molecule plays no role in the specified dynamics, the analysis of sections 2 and 3 leads to the assignment of properties to the molecule as a whole. Within this framework the molecule as a whole is an atomic system, and formula (3.1) provides a consistent joint probability distribution for properties on this level of detail. Alternatively, if the dynamics is in terms of nucleons and electrons, *these* systems should be considered as atomic; again we can write down a joint probability distribution. We are thus led to the point of view that atomic systems are the blocks from which systems are built up; but it depends on the level of description (the coarseness of the specified dynamics) *which* systems are identified as atomic.

Another illustration of the same idea is the following variation on the two-slit experiment. Suppose that a two-particle system can pass through two holes in a screen. A possible quantum state for this situation is

$$|\Psi\rangle = a|\psi^{\alpha}\rangle\,|\psi^{\beta}\rangle + b|\phi^{\alpha}\rangle\,|\phi^{\beta}\rangle \tag{4.1}$$

where the first term on the right-hand side refers to states (of particles α and β, respectively) that are localized in the first slit and the second term to states localized in the second slit (these states are orthogonal to the states occurring in the first term). The state in equation (4.1) has the form of a biorthogonal decomposition, so according to the property attribution rule both α and β are localized (with probability $|a|^2$ in slit 1 and with probability $|b|^2$ in slit 2). Measurements that are specific enough to probe the individual particles will confirm this: no interference will be found between, for instance, particle α going through slit 1 and it going through slit 2. If the interaction with the outside world occurs through some collective coordinate, however, interference effects *will* show up. If the interaction is not sensitive to the two-particle character of the system at all, everything will appear to the outside world as if one nonlocalized particle passes the screen. Clearly, the descriptions on different levels of detail can be more drastically different in quantum mechanics than in classical physics: the relevant definite observables will in general fail to commute.

The Bohm-EPR case is an obvious further case to consider. As before, interactions with outside systems will lead to the same results (Dieks 1994), regardless of whether emergent properties are attributed to the singlet system or not. In this example the individual spin systems are attributed trivial properties, represented by a multiple of the unity operators in the two-dimensional spin Hilbert spaces (because of the degeneracy in the total singlet state). Obviously, these operators commute with all other operators in the spin part of the total Hilbert space. No harm would therefore be done if the projector on the total singlet state would be inserted in the probability formula equation (3.1).

This inspires a possible slight adaptation of our suggested strategy, namely, to include in the property ascription the properties given by the biorthogonal decomposition rule to nonatomic systems, if the associated projectors commute with the definite-valued projectors assigned to the atomic systems. It could be argued that this has the advantage of providing an explanation, on the level of possessed properties, for the correlation between the outcomes of measurements on the individual spin systems (Healey 1989): although the individual spins do not have definite directions themselves, the total spin has the definite value zero.[2]

5. Macroscopic Properties and Decoherence

We have indicated how in an interaction atomic systems may acquire properties that reflect a collective effect of a number of other systems. In real measurements, however, macroscopic devices sometimes register such collective quantum effects. We should therefore be able to explain how properties of atomic systems can be transferred to the level of macroscopic properties. More generally, the question to be considered should be how macroscopic properties can be explained in terms of properties of atomic constituents.

The latter question has already received considerable attention within "nonatomic" versions of the modal interpretation (Bacciagaluppi and Hemmo 1996). It has turned out that decoherence has to be taken into account in order to make it understandable that the usual physical magnitudes, known from everyday experience, are definite-valued in macroscopic objects (or, rather, that observables which are extremely close to those "classical" ones are definite). The details of these calculations have to be reconsidered for the "atomic version" of the modal interpretation, something that is outside the scope of the present chapter. We shall confine ourselves here to two remarks.

First, the atomic version which was outlined in the foregoing offers prospects of treating macroscopic properties more or less along the lines of classical physics. In classical physics the properties of macroscopic bodies supervene on the properties of their atomic constituents; the same is true in the new version of the modal interpretation. We can therefore adopt strategies similar to those in classical physics to make it understandable that microscopic properties can be transferred to the macrolevel.

The paradigmatic classical case of a micro-macro transition is that in which particles in a rigid body perform a collective motion. Think, for example, of the motion of the pointer of a measuring device. Because of the rigid connections between the particles, the pointer as a whole essentially executes the same motion as its constituents. An important point of difference between classical physics and quantum mechanics that makes it impossible in general to simply copy the above for the case of quantum mechanics, is that in classical physics position always is a possessed property whereas in quantum mechanics *any* observable can in principle become definite-valued, thereby excluding the definiteness of all observables that do not commute with it. Position has consequently no a priori preferred status and need not be definite (at least, that is the situation according to the version of the modal interpretation we are discussing). However, the definiteness of position (or something very close to it) is crucial, not only for the functioning of measuring instruments but for the transition from quantum mechanics to classical physics in general. This is one point at which the mechanism of decoherence appears to be of vital importance.

To bring decoherence into play, we assume that the Hamiltonian of our system contains terms representing an interaction with the environment (which is assumed to possess very many degrees of freedom). As we have seen, the Hamiltonian defines a level of description in terms of atomic systems. It is well known that under conditions typically assumed in model calculations the effect of the interaction with the environment will be to select position (or some observable extremely close to position) as a definite-valued observable for systems that are in interaction with the environment (one can think here of those atomic constituents that are part of the "surface layers" of the macroscopic object). If it can indeed be borne out by a detailed treatment that those atomic constituents of a macroscopic body which are in contact with the environment acquire definite positions, this would be a decisive step toward solving the problem of why position can always be assumed to be definite on the macroscopic level. Indeed, property composition will ensure that the macroscopic system has a well-defined spatial boundary, demarcating it from the environment.

A second aspect to the process of decoherence seems equally important to understand the transition to the classical regime. Decoherence appears responsible for the fact that there are no collective quantum effects on the classical level. Consider the simple case of a system consisting of two parts, $\alpha\&\beta$. Suppose that the system is decohered by its environment in the following way:

$$|\Psi\rangle = \sum_{i,j} c_{ij} \, |\psi_i^\alpha\rangle \, |\psi_j^\beta\rangle \, |\psi_{ij}^E\rangle, \tag{5.1}$$

with $\langle\psi_k^\alpha|\psi_l^\alpha\rangle = \langle\psi_k^\beta|\psi_l^\beta\rangle = \delta_{kl}$ and $\langle\psi_{ij}^E|\psi_{kl}^E\rangle = \delta_{ik}\delta_{jl}$; the superscript E refers to the environment. We assume here that the two parts of the system are individually and independently decohered by the environment. This looks like a natural assumption for small systems in contact with an environment containing a very large number of degrees of freedom. Application of the biorthogonal decomposition rule now yields that $|\psi_i^\alpha\rangle\langle\psi_i^\alpha|$ and $|\psi_j^\beta\rangle\langle\psi_j^\beta|$ are the definite-valued projectors of α and β, respectively. Application of the same rule to the compound system $\alpha\&\beta$ selects $|\psi_i^\alpha\rangle\langle\psi_i^\alpha| \otimes |\psi_j^\beta\rangle\langle\psi_j^\beta|$ as being definite. In other words, if we were to apply the biorthogonal decomposition rule directly to the composite system $\alpha\&\beta$, this would not lead to any new properties not found by property composition. The motivation to introduce "emergent" properties disappears in such circumstances. For our suggested atomic approach, in which there are no emergent properties anyway, it is important that if decoherence takes place in the manner of equation (5.1), interactions with $\alpha\&\beta$ will not lead to results that reflect collective effects. This is because whatever the interaction is between some third system, θ, and $\alpha\&\beta$, the probability distribution of properties of θ will not show any contribution of interference between different pairs of indices $\{i,j\}$. Interference terms vanish in the calculation because of the presence of the mutually orthogonal environmental states $|\psi_{ij}^E\rangle$.

Summing up, it seems that if the effects of decoherence are taken into account, the atomic version of the modal interpretation offers prospects for understanding the transition to the level where the usual macroscopic physics applies.

6. Conclusion

We have explored, in a rather tentative and programmatic way, possible solutions to open problems in the modal interpretation of quantum mechanics. First, it seems possible to exclude factorizations rival to a given physical one if an embedding in space-time is given (by means of a representation of space-time transformations including a Hamiltonian containing nontrivial interaction terms). Given this unique physical factorization, a simple expression for the joint probability distribution of possessed properties can be given if it is assumed that all properties follow by property composition from atomic properties.

According to the point of view developed here, the atomicity of systems is relative to the "coarseness" of the dynamical description (the amount of detail in the Hamiltonian). Although there are no emergent properties in this view, there may be typically quantum-mechanical collective effects on other systems. We have argued, however, that it is plausible that in the presence of decoherence the behavior of systems becomes classical in important respects.

Notes

I am indebted to Guido Bacciagaluppi, Tim Budden, and Pieter Vermaas for helpful discussions.

1. Guido Bacciagaluppi has pointed out (private communication) that a consideration of more than two particles makes it possible to strengthen the argument, so that the value of G no longer plays a role.

2. The value of this addition to the set of definite properties is questionable, however. The suggested explanation will not generalize to nondegenerate variants of the Bohm-EPR experiment. Moreover, it is not clear exactly how emergent properties are supposed to fulfill their explanatory function here. The very fact that emergent properties do not supervene on properties of the individual systems makes it hard to see how they could "regulate" the outcomes of measurements on these individual systems, and thus be responsible for correlations in the measurement results. It seems that even if emergent properties are accepted, some form of global dynamics is needed.

References

Bacciagaluppi, G. 1995. "Kochen-Specker Theorem in the Modal Interpretation of Quantum Mechanics." *International Journal of Theoretical Physics* 34: 1206–15.

Bacciagaluppi, G., and W. M. Dickson. 1995. "Modal Interpretations with Dynamics." Preprint, Cambridge University. Also, "Dynamics for Density Operator Interpretations of Quantum Theory." 1997. quant-ph/9711048 (available in early 1998 on the Quantum Physics electronic archive at http://xxx.lanl.gov).

Bacciagaluppi, G., and M. Hemmo. 1996. "Modal Interpretations, Decoherence, and Measurements." *Studies in the History and Philosophy of Modern Physics* 27: 239–77; see also the chapter by the same authors in this volume.

Clifton, R. 1995. "Why Modal Interpretations of Quantum Mechanics Must Abandon Classical Reasoning about the Values of Observables." *International Journal of Theoretical Physics* 34: 1303–12.

Dieks, D. 1989. "Quantum Mechanics without the Projection Postulate and Its Realistic Interpretation." *Foundations of Physics* 19: 1397–1423.

———. 1994. "Modal Interpretation of Quantum Mechanics, Measurements, and Macroscopic Behavior." *Physical Review* A 49: 2290–2300.

———. 1995. "Physical Motivation of the Modal Interpretation of Quantum Mechanics." *Physics Letters* A 197: 367–71.

Feynman, P. F., R. B. Leighton, and M. Sands. 1965. *The Feynman Lectures on Physics.* Vol. 3, chapter 9. Reading, Mass.: Addison-Wesley.
Healey, R. 1989. *The Philosophy of Quantum Mechanics: An Interactive Interpretation.* Cambridge: Cambridge University Press.
Vermaas, P. E., and D. Dieks. 1995. "The Modal Interpretation of Quantum Mechanics and Its Generalization to Density Operators." *Foundations of Physics* 25: 145–58.

Michael Dickson

On the Plurality of Dynamics: Transition Probabilities and Modal Interpretations

1. Review of Modal Interpretations

1.1. The Basic Idea

There are by now many children and grandchildren of van Fraassen's (1972) modal interpretation of quantum mechanics.[1] These interpretations share the same central idea: quantum-mechanical systems can possess properties even if the state of the system does not confer probability 1 on that property. The usual slogan is that modal interpretations "deny the eigenstate-eigenvalue link," which means that a system can possess the value a for the observable A without being in the eigenstate $|a\rangle$ (where $A|a\rangle = a|a\rangle$). In the terms that will be used here, a system can possess the property corresponding to the projection operator $|a\rangle\langle a|$ ($=$ df $P_{|a\rangle}$) even when it is not in the state $|a\rangle$. (Assume a generic one-to-one correspondence between projection operators on the Hilbert space of a system and properties of that system.)[2]

The trick, of course, is to say *which* properties a system can possess. One strategy (suggested, for example, by Vink [1993]) is to take *every* property to be definite-valued (that is, either possessed or not), so that the algebra of definite-valued properties for a system is the lattice of projections on the system's Hilbert space. However, this strategy leads to a contradiction (via the Kochen-Specker theorem), which is avoided only by a subtle daring not approved by most advocates of modal interpretations. Hence they must pare down the lattice of projections on Hilbert space to some set (hopefully also an algebra under some suitable operations) that can consistently and straightforwardly be said to be the set (or algebra, under some suitable algebraic operations) of definite-valued properties. For short, call these properties the "definite properties." Each definite property is either possessed or not by a system, whereas all other "properties" are neither possessed nor not possessed.

Suppose that the system of interest is in the quantum-mechanical state (that is, has the density operator) W. Allowing that the set of definite properties may depend on W, call the set of definite properties A_W. (A_W may also depend on other things, but it depends on W in every extant modal interpretation.) A_W should meet two basic constraints. First, it should contain those properties that we believe on independent grounds to be definite (for example, the train's leaving at roughly two o'clock, the coffee's being hot, the picture's being roughly in the middle of the wall, and so on), or at the very least it should contain the properties that you and I have when we believe such properties to be definite. Second, the theory based on A_W should be empirically equivalent to standard quantum mechanics. I shall not discuss the first constraint any further here.[3] The most straightforward way to meet the second constraint is to stipulate that the probability that any given property in A_W is possessed by a system is the quantum-mechanical one; in other words, for any $P \in A_W$, stipulate that the probability that P is possessed by a system whose state is W is $\mathrm{Tr}[WP]$.

Different modal interpretations choose different sets of definite properties, but they are unified in beginning with an algebra of properties that may be called a "faux-Boolean algebra."[4] (Because of further modifications of the algebra, as discussed in the next two subsections, not all modal interpretations *end up* with a faux-Boolean algebra, however.) Such algebras are constructed as follows. Take some set, S, of mutually orthogonal projections on some Hilbert space, H. Add to it every projection in the subspace orthogonal to the subspace spanned by S. Close under the lattice-theoretic operations (infimum, supremum, and orthocomplement—or meet, join, and not). The resulting structure (which is a lattice, but not in general a Boolean one) is a "faux-Boolean algebra," B.

A simple example is depicted in Figure 1 for a real, three-dimensional Hilbert space where S contains just a single element. The atoms of this faux-Boolean algebra are the single element of S plus all rays orthogonal to it. By closure, all planes spanned by two atoms are added (some of these are depicted), as well as the unit projection.

These algebras are called "faux-Boolean" because despite being in general non-Boolean, they can all be said to have a classical model in the following sense. Given any set of events, C, any map, $\mathrm{p}: C \rightarrow [0,1]$, and any map $\mathrm{p}_{\mathrm{cond}}: C \times C \rightarrow [0,1]$, the "theory" $\langle \mathrm{p}, \mathrm{p}_{\mathrm{cond}}, C \rangle$ has a *classical model* if and only if there exists some classical theory, $\langle \mathrm{p}_{\mathrm{cl}}, \mathrm{p}_{\mathrm{cl\ cond}}, F \rangle$ (where p_{cl} is a classical probability measure on F, $\mathrm{p}_{\mathrm{cl\ cond}}$ is a classical conditional probability measure on $F \times F$, and F is a Boolean algebra), and there exists some map, $\mu: C \rightarrow F$, such that for all C_1, C_2 in C:

1. $\mathrm{p}[C_1] = \mathrm{p}_{\mathrm{cl}}[\mu(C_1)]$
2. $\mathrm{p}_{\mathrm{cond}}[C_1, C_2] = \mathrm{p}_{\mathrm{cl\ cond}}[\mu(C_1), \mu(C_2)] \left(= \dfrac{\mathrm{p}_{\mathrm{cl}}[\mu(C_1) \cap \mu(C_2)]}{\mathrm{p}_{\mathrm{cl}}[\mu(C_2)]} \right).$

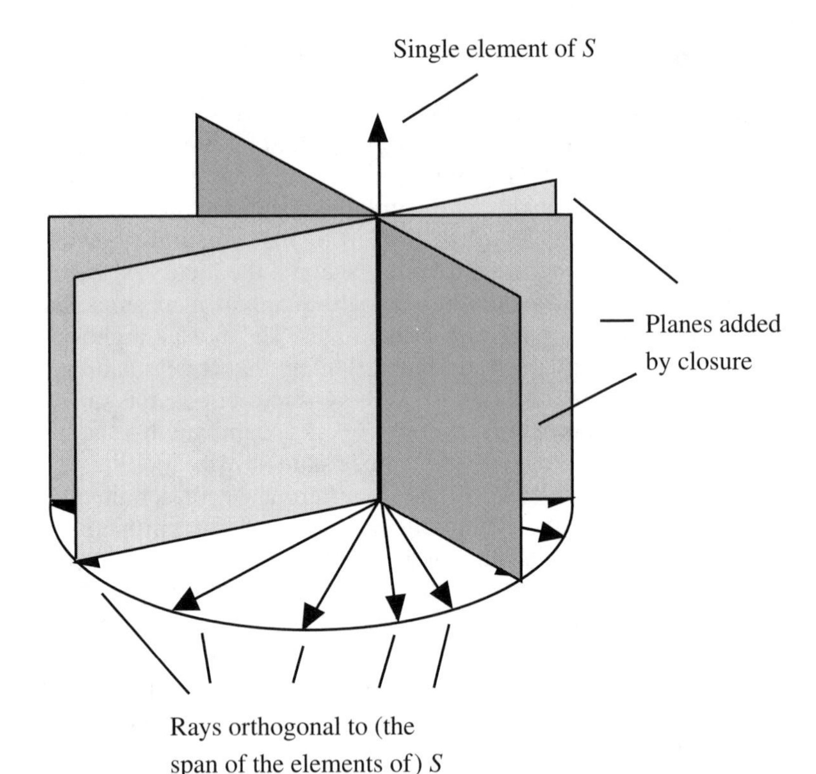

Figure 1. Example of a faux-Boolean algebra in a real, three-dimensional Hilbert space.

For a certain class of "faux-classical" probability measures, p_{fc} and $\mathrm{p}_{\mathrm{fc\,cond}}$, over any given faux-Boolean algebra, B, the theory $\langle \mathrm{p}_{\mathrm{fc}}, \mathrm{p}_{\mathrm{fc\,cond}}, B\rangle$ always has a classical model. (See Dickson [1995b] for a proof.) For a given faux-Boolean algebra, B, faux-classical measures are those measures over B that assign all the probability to the elements of the set S used to generate B, that is,

$$\sum_{P_i \in S} \mathrm{p}_{\mathrm{fc}}[P_i] = 1 \tag{1}$$

and zero probability to all projections orthogonal to the span of the projections in S.[5]

As it turns out, given the algebra A_W as chosen by any extant modal interpretation in which A_W is faux-Boolean, the measure generated by W

is a faux-classical measure, so that modal interpretations are empirically classical. In other words, even though the algebra A_W is in general non-Boolean, the empirical probabilities (which here are just the probabilities that a system possesses one of the properties in A_W, not the probabilities that one of these properties will be observed) will be such as could have come from a classical probability space. In these theories, the nonclassicality of the world is hidden. (A similar story can be told about the logic of modal interpretations [Dickson 1996].)

Indeed, more can be shown. A faux-Boolean algebra is *maximally* classical in the following sense. Call any projection that is contained in some atom of B a "subatomic" projection. Suppose that some nonsubatomic projection is added to a faux-Boolean algebra, B, and that the resulting set is then closed under the lattice-theoretic operations, yielding a new algebra of properties, B_{plus}. The algebra B_{plus} has no classical model (Dickson 1995c). Therefore, modal interpretations in a sense choose the largest set they can while still having a classical model. (One must say "in a sense" because subatomic properties can be added. Moreover, the theorem cited above does *not* mean that there is not some "larger" structure—given some appropriate notion of "larger"—that has a classical model. Rather, the theorem says that nothing except subatomic properties can be added to a faux-Boolean algebra and still have a classical model. Note also that the addition of a [single] subatomic property produces a new faux-Boolean algebra, thus guaranteeing the existence of a classical model.)

1.2. Examples

Although the details of various modal interpretations are not crucial here, it might help to show a few examples. Bub's interpretation begins by specifying some privileged observable, R, which is always definite-valued. For a system in a pure state, W_{pure}, Bub's algebra of definite properties is then constructed by letting the set S be the set of projections of W_{pure} onto the eigenspaces of R. Bub (1992, 1994) has proposed a corresponding construction for mixed states, but a recent discussion by Bub and Clifton (1996) suggests that one need only perform the construction for the state of the universe (which is presumably pure), and then derive the set of definite properties for subsystems (see the next subsection).

An interpretation roughly attributable to Kochen (1985), Dieks (1988, 1989), and Healey (1989) begins with the biorthogonal decomposition theorem (or, for Kochen, the polar decomposition theorem [Reed and Simon 1975, 196–98]). Suppose that a system has the statevector ψ in a Hilbert space $H^{\alpha\beta} = H^{\alpha} \otimes H^{\beta}$, where H^{α} and H^{β} are the Hilbert spaces for subsystems α and β. The biorthogonal decomposition theorem says that the decomposition

$$\psi = \sum_i c_i(a_i \otimes b_i)$$

exists (where $\{a_i\}$ is an orthonormal set in H^α and likewise $\{b_i\}$ in H^β), and is unique so long as $|c_i| = |c_j|$ implies that $i = j$. According to this interpretation, then, whenever the biorthogonal decomposition is unique, a (faux-Boolean) algebra of properties for the subsystem α is constructed by letting the set S be the set of projections onto the vectors $\{a_i\}$, and likewise for β. Authors differ on what to do in the case of nonuniqueness, and some authors, especially Healey, make several alterations to the basic faux-Boolean algebra of properties.

The interpretation of Vermaas and Dieks (1995) is a generalization of the above. In it, the set of definite properties is constructed by letting S be the set of projections in the (unique) spectral resolution of a system's density operator, W. (The unique spectral resolution is the set of projections appearing in the spectral decomposition of W, which exists and is unique by the spectral decomposition theorem, although some of the projections in it may be more than one-dimensional.) In the case where two systems, α and β, are subsystems of a larger system in a pure state, ψ, with a unique biorthogonal decomposition, this interpretation implies the previous one.

1.3. Compound Systems

The previous subsections focused on single systems. The examples from section 1.2 show how some modal interpretations differ in their selection of A_W for single systems. They may also differ in their treatment of compound systems. The basic differences concern four points: the definition of a subsystem, property composition, property decomposition, and supervenience. A detailed account of the position of each existing modal interpretation on these points is out of the question. Below is a brief discussion and a few examples.[6]

Consider a system composed of two subsystems, α and β. The Hilbert space for the compound system is $H^{\alpha\beta} = H^\alpha \otimes H^\beta$. A generic property for α (projection onto H^α) is denoted P^α, and likewise for P^β.

The problem of defining a subsystem arises by considering a second factorization of $H^{\alpha\beta}$ into $H^{\alpha'}$ and $H^{\beta'}$. For the sake of emphasizing the problem, suppose that $\dim(H^\alpha) = \dim(H^{\alpha'})$ and $\dim(H^\beta) = \dim(H^{\beta'})$, so that the primed factorization is just a "rotation" of the unprimed one. We may assume that subsystems α and β receive their own algebras of definite properties according to the procedure of some modal interpretation. The question now is whether α' and β' do also. Not all advocates of modal interpretations have considered this question, but among those who have, Dieks has taken the view that every factorization determines a subsystem,[7] while Healey takes the view that there is a "preferred" factorization that determines what the subsystems are. Neither approach is very attractive, but the latter seems clearly the better, though it faces the difficult problem of justifying one factorization as the one that picks out the "real" subsystems.

Property composition is the condition that if α possesses P^α and β possesses P^β, then the compound system $\alpha\&\beta$ possesses $P^\alpha \otimes P^\beta$. Property decomposition is the condition that if $\alpha\&\beta$ possesses $P^\alpha \otimes P^\beta$, then α possesses P^α and β possesses P^β. Supervenience is the condition that a compound system possess only properties that are products of properties of subsystems or algebraic combinations of such properties. That is, supposing that $\{P_i^\alpha\}$ and $\{P_j^\beta\}$ are the definite properties for α and β, $\alpha\&\beta$ can have as definite properties only those in the lattice-theoretic closure of $\{P_i^\alpha \otimes P_j^\beta\}$. Some or all of these conditions are denied by various modal interpretations. (Note that the interpretations as given in section 1.2 concerned single systems only; in theory they are all free to accept or deny any of the conditions given above.)

For example, as already mentioned, the discussion of Bub and Clifton (1996) suggests that an algebra of definite properties is given first of all for the universe (presumed to be in a pure state). Properties are then assigned to subsystems through the condition that P^α is definite for α if and only if $P^\alpha \otimes 1$ (a projection on the Hilbert space of the universe) is definite for the universe. (They do not say explicitly whether every factorization represents a subsystem.) Property composition, decomposition, and supervenience appear to hold in Bub's interpretation.

Healey imposes a number of conditions on algebras of events for subsystems, but it remains unclear what the consequences of these conditions are for Healey's set of definite properties. It appears that this set supports property composition and property decomposition, but not supervenience.

Dieks (1988, 1989), as well as Vermaas and Dieks (1995), appear to deny all three. They treat every system on its own, deriving an algebra of definite properties from the system's density operator. Consider, for example, the density operator $W^{\alpha\beta}$, and the reduced density operators (obtained from $W^{\alpha\beta}$ by partial tracing) W^α and W^β. Clearly the eigenprojections of $W^{\alpha\beta}$ need not be products of eigenprojections of W^α and W^β. Therefore, because the compound system $\alpha\&\beta$ gets all of its properties by applying Vermaas and Dieks's procedure to $W^{\alpha\beta}$, property composition fails. Perhaps less obviously, property decomposition fails as well. To see why, consider a case where

$$W^{\alpha\beta} = w_1 P_1^\alpha \otimes P_1^\beta + w_2 P_2^\alpha \otimes P_2^\beta$$

and where P_1^β is orthogonal to P_2^β but P_1^α is not orthogonal to P_2^α. In this case, the reduced state of α is $W^\alpha = w_1 P_1^\alpha + w_2 P_2^\alpha$, but because P_1^α and P_2^α are not orthogonal, this expression does not give the spectral resolution of W^α. Therefore, the compound system $\alpha\&\beta$ could possess $P_1^\alpha \otimes P_1^\beta$ while α fails to possess P_1^α, and property decomposition fails. Supervenience clearly fails as well.

As a final example, Bacciagaluppi and Dickson (1998) have proposed an interpretation according to which faux-Boolean algebras are assigned (along the lines advocated by Vermaas and Dieks [1995]) *only* to a chosen set of "atomic" systems (given by some preferred factorization); all properties of compound systems are derived from the properties of the atomic systems via property composition (and algebraic combinations of such properties). In this proposal, property composition of course holds, as do property decomposition and supervenience.

Note that the result of some of the choices made by some modal interpretations is that the final algebra of definite properties for some systems will *not* be a faux-Boolean algebra. In the case of Dieks (1988, 1989) and Vermaas and Dieks (1995), this fact is obvious, because they assign properties to a subsystem based on *any* factorization of the Hilbert space for the universe of which it is a part. In the case of Healey, the conditions imposed appear to result in a set of definite properties that is not even a partial Boolean algebra,[8] much less a faux-Boolean algebra. On the other hand, the interpretation of Bub and Clifton and the proposal of Bacciagaluppi and Dickson yield algebras for any subsystem that *is* a faux-Boolean algebra.

A great deal more could be said about the differences among the various modal interpretations, but this subsection and the previous two contain enough detail to give a reasonable indication of the general program. I turn now to the problem of dynamics.

2. Dynamics[9]

2.1. What We Need

Modal interpretations as I have described them thus far can at best specify at any time and for any system a set of definite properties and a probability measure on that set. What is clearly lacking is an account of how a system's properties change over time.

One's attitude toward this lack of a dynamics will depend in part on one's attitude toward the project of interpretation in the first place. If the aim is merely to put quantum mechanics—as far as it goes—on a consistent footing, then perhaps a dynamical story is not necessary. On this view, what quantum mechanics is lacking is just a "preferred basis," or more generally, a set of definite properties. Given such a set, the meaning of any quantum-mechanical state is evident: it is a probability distribution over the definite properties.

Once one has postulated that a given set of properties is "preferred" in this way, that is, that a given set of properties is *the* set of physically definite properties, then it is at least natural—some might say compelling—to ask the question, "How do a given system's definite properties change over time?" If one takes the aim of interpretation to be to say how the world is, or, in a more antirealistic spirit, what the best model of empirical phenomena is, then it seems all the more crucial to answer this question.

It might appear at first that this question, or at least the answer to it, would look different in different modal interpretations, but there is a way to answer it for any modal interpretation whose set of definite properties is at each time a faux-Boolean algebra, B, and for which the probability measure over B is faux-classical. Recall that in these cases, equation (1) holds. Now assume that one and only one element in S is possessed at any given time, and that if some atomic property, P_k, is possessed, then all and only the consequences of P_k that are in B are also possessed. Then the problem of finding a dynamics for the possession of properties in B is reduced to the problem of finding a dynamics for the possession of properties in S.

However, the move to a dynamics for compound systems is not always as straightforward. In section 1.3 we saw why: some modal interpretations end up with algebras of definite properties that are not faux-Boolean algebras, and not even necessarily partial Boolean algebras. There are *conditional* proposals in place (Bacciagaluppi and Dickson 1998), showing how a complete dynamics *could* be found *if* these interpretations are shown to have well-defined joint probabilities for compound systems, but this "if" seems very doubtful at the moment.

On the other hand, some modal interpretations can make the move to compound systems quite simply in the case of single-time probabilities, and equally so for a dynamics. In the case of Bub and Clifton, a dynamics for a single system (the universe) induces a dynamics on all subsystems in a well-defined way. The situation is similar in the case of Bacciagaluppi and Dickson: a dynamics for the universal system—a composite system made up of all of the atomic systems in the universe—induces a dynamics for any subsystem of the universal system in a well-defined way. (See Bacciagaluppi and Dickson [1998] for details. This procedure requires joint probabilities for possession of properties by two or more atomic systems; these probabilities were given by Vermaas and Dieks [1995].) Therefore, a dynamics for single systems only is given here.

2.2. Preparatory Discussion

Denote the probability that the projection $P_i \in S$ is possessed by "p_i." Denote the probability that P_j is possessed at time $t + dt$ given that P_i was possessed at t by "$\mathrm{p}_{j|i}$." We want a complete set of such $\mathrm{p}_{j|i}$, and we require of them that they return the faux-classical measure (which is of course also the quantum-mechanical measure) as a marginal. That is, $\Sigma_i\, \mathrm{p}_{j|i}\, \mathrm{p}_i = \mathrm{p}_j$.

One way to get what we need is trivial: let $\mathrm{p}_{j|i} = \mathrm{p}_j$. This measure yields a dynamics, but one in which a system's properties at one time are completely uncorrelated with its properties at a later time. A strange sort of dynamics, but one that is apparently empirically adequate.[10]

If we want more than this completely uncorrelated dynamics, it helps to know something about how the set S evolves in time. In some modal interpretations (for example, some versions of Bub and Clifton's interpretation), S is effectively constant.[11] In others (for example, the proposal of Bacciagaluppi and Dickson), S is time-dependent, but evolves continuously (in the sense that its eigenspaces evolve continuously through Hilbert space).[12] In both cases, it is therefore easy to set up a one-to-one correspondence between the elements of S at time t and the elements of (a possibly new) S at time $t + dt$.

In principle one can simply choose any correspondence one likes. The only apparent obstacle is that S may have different cardinalities at different times. Yet there is really no problem here. We can simply find the $S(t)$ (S at time t) with the greatest cardinality, then decompose the elements of $S(t)$ at the other times into "fiduciary" elements. For example, supposing that the set of greatest cardinality has cardinality N, while another set, at time s, has cardinality $N - 2$. In that case, there must be some multidimensional projections in either $S(s)$ or $S^{\perp}(s)$ (because the dimension of a system's Hilbert space does not change in time). Suppose for illustration that $S(s)$ contains a three-dimensional element, P. Then we can decompose P into three one-dimensional, mutually orthogonal, projections, $P = \tilde{P}_1 + \tilde{P}_2 + \tilde{P}_3$, replacing P in $S(s)$ with these three projections. The resulting faux-Boolean algebra will contain the old one as a subalgebra. Hence any dynamics involving the new algebra will induce a dynamics on the old algebra. (For example, if any of the $\tilde{P}_i$ is possessed, then we will say that P is possessed.) In this way, we can always define *some* one-to-one correspondence among elements of $S(t)$ at any different times.

Doing so creates a set of (not necessarily continuous) "trajectories" through Hilbert space, and reduces our problem to one of finding the probability at any time of a transition from one trajectory to another. (A trajectory is defined by an initial projection plus its one-to-one correspondents at all later times.) In other words, we may imagine a system's possessed property to travel along these trajectories, occasionally making a jump from one trajectory to another. What we need, finally, is the probability of jumping from any one trajectory to any other at any given time. From now on, let the indices i and j refer to entire trajectories, rather than to single projections.

2.3. How to Get What We Need

Now for some details. Define the probability current J_{ji} to be the net flow of probability from i to j. (J_{ji} is a current and may be conceived analogously to the current in fluid dynamics.) Define a transition probability matrix, T_{ji}, such that $T_{ji}dt = \mathrm{p}_{j|i}$. (Here I follow Vink [1993]. See also Bacciagaluppi and Dickson [1998].) Given these definitions, we have

$$J_{ji} = T_{ji}\mathrm{p}_i - T_{ij}\mathrm{p}_j, \tag{2}$$

which says that the net flow of probability from i to j is the flow into j from i minus the flow out of j into i. Clearly J_{ji} must be antisymmetric. A further restriction on J_{ji} is that it obey the continuity equation:

$$\sum_i J_{ji} = \dot{\mathrm{p}}_j. \tag{3}$$

If we can find expressions for J and T that satisfy (2) and (3), then our problem is solved. As I will discuss later, there are many such expressions. Here, I will give just one solution for T in terms of J, due to Bell (1987), and two solutions for J, due to Bacciagaluppi and Dickson (1998).

Bell's choice for T is motivated by the Bohm theory, in which the velocity of a particle is given by the current divided by the density. (See Vink [1993] for further discussion of the relation of the following to Bohm's theory.) Hence, for $i \neq j$, Bell chooses[13]

$$T_{ji} = \max(0, J_{ji} / \mathrm{p}_i) \tag{4}$$

Note: the antisymmetry of J_{ji} makes (4) a solution of (2). The probability to stay in the same state, $T_{ii}dt$, follows from normalization:

$$\sum_j T_{ji}dt = 1. \tag{5}$$

For the current, J, there is a standard expression found in textbooks, which is readily derived from the Schrödinger equation (though it is usually given only in the position representation, rather than in the "S-representation"; see, for example, Schiff [1955, 23–24]):

$$J_{ji}(t) = 2\ \mathrm{Im}[\langle \psi(t)|\mathrm{P}_j H \mathrm{P}_i|\psi(t)\rangle], \tag{6}$$

where P_i and P_j are elements of S, H is the Hamiltonian of the system and ψ is its statevector. However, (6) does not apply to cases where S changes in time. In this case, (6) must be generalized. To begin, note that

$$\dot{p}_j(t) = 2\ \mathrm{Im}[\langle \psi(t)|P_j(\mathrm{t})H|\psi(t)\rangle] + \langle \psi(t)|\dot{P}_j(\mathrm{t})H|\psi(t)\rangle \tag{7}$$

where $\dot{P}_i$ is the time derivative of P_j. To generalize (6), we need to find a current that satisfies (3) and (7). There is no *unique* way to do so. Perhaps the most obvious way is to choose

$$J_{ji}(t) = 2\ \mathrm{Im}[\langle \psi(t)|\mathrm{P}_j(\mathrm{t})H\mathrm{P}_i(\mathrm{t})|\psi(t)\rangle] + \langle \psi(t)|\dot{P}_i(\mathrm{t}) - \dot{P}_j(\mathrm{t})|\psi(t)\rangle \tag{8}$$

but the following expression, for example, is also satisfactory:

$$\begin{aligned} J_{ji}(t) = {} & 2\ \mathrm{Im}[\langle \psi(t)|P_j(t)HP_i(t)|\psi(t)\rangle] \\ & + \mathrm{Re}[\langle \psi(t)|(\dot{P}_j(t)P_i(t) - \dot{P}_i(t)P_j(t))|\psi(t)\rangle]. \end{aligned} \tag{9}$$

Equations (4) and (8), or (4) and (9), are sufficient to specify a complete dynamics.

3. A Plurality of Dynamics

3.1. Why There Is a Plurality

In the previous section, we saw two ways to solve equations (2) and (3), but there are many more. This point is easy enough to see in both (2) and (3). Equation (2) is underdetermined. As Vink (1993) has noted, one can add to the T_{ji} defined in (4) any solution, T^0, of the equation

$$T^0_{ji}\mathrm{p}_i - T^0_{ij}\mathrm{p}_j = 0. \tag{10}$$

The result is an enormous freedom in the choice of a solution to (2).

Likewise, equation (3) is underdetermined. In standard quantum mechanics, we think only of using the usual Schrödinger current, as given in equation (6), but apart from its usefulness in certain contexts, nothing in particular recommends it as an expression for the current—*any* solution of (3) will apparently yield an empirically adequate dynamics. Furthermore, (6) is not applicable to the case where S is time-dependent; in this case one must generalize the Schrödinger current, as in equations (8) and (9), and although these are certainly natural ways to generalize (6), they are not the only ways. As was the case for T, given any expression for J that solves (3), a new expression can be constructed by adding to J a solution, J^0, of the equation

$$\sum_i J^0_{ji} = 0. \tag{11}$$

There is no obvious criterion to select just one of these possible solutions.

Indeed, as will be discussed below, different contexts might make different choices for the current more useful, or natural. A principle of conservatism might dictate that one choose a current of "minimal flow"—one, for example, that minimizes the quantity $\Sigma_{i,j} J^2_{ji}$. (See Bacciagaluppi and Dickson [1998] for a dynamics that does so.) As long as the chosen current is a solution of (3), it will be empirically satisfactory, in the sense that it will yield probabilities that match those of quantum mechanics.

Therefore, using the flexibility in choosing solutions of (2) and (3), there are many ways to find a dynamics. Whether, for a given modal interpretation, these many ways correspond to all possible dynamics is an interesting mathematical question (to which I believe the answer is yes but have no proof), but more important here is the mere fact that there are already many empirically adequate dynamics. Apparently, there is no decisive way to choose among them.

3.2. Is the Plurality Genuine?

Or is there? Although there is no available empirical criterion for deciding which among many dynamics is *the* dynamics, there might be other criteria that may be brought to bear. Consider, for example, the criterion of locality. We know from the no-signaling theorems that a minimal requirement from the theory of relativity—namely, the absence of superluminal signaling—will be met by any theory that is empirically equivalent to standard quantum mechanics. Therefore, no *empirical* criterion can be brought to bear from the theory of relativity, but some may find in the theory of relativity the motivation for a slightly stronger form of locality. To see how such a motivation might be relevant to a choice of dynamics, let us consider a standard EPR-Bohm experiment, as depicted in Figure 2.

The particles begin in the singlet state at time t_0 (in some frame), and in a space-time region beginning at time t_m the spin of each particle is measured in the z-direction. Suppose that in some modal interpretation, at time t_0, particle 1 possesses either $P_{|+,z\rangle}$ (the property of having spin-up in the z-direction) or $P_{|-,z\rangle}$. Because the quantum-mechanical joint probabilities are always obeyed in a modal interpretation, one particle will possess $P_{|+,z\rangle}$

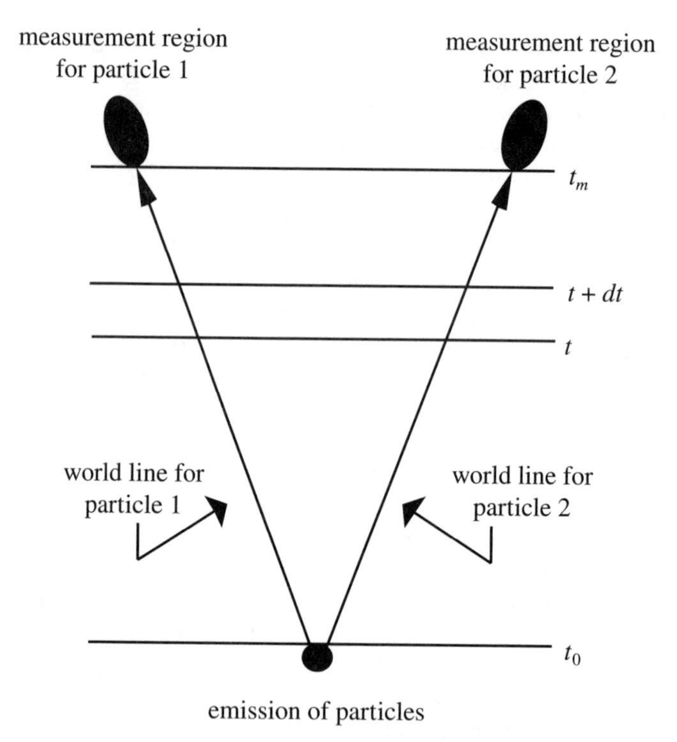

Figure 2. Space-time diagram of a standard EPR-Bohm experiment.

and the other will possess $P_{|-,z\rangle}$. Finally, suppose that in our modal interpretation, the set of definite properties does not change between t_0 and t_m.

Now, if we choose a dynamics in which no transition occurs between t_0 and t_m, then the result of the experiment can be explained completely in terms of the initial states of the particles, and not a whiff of nonlocality need disturb the air. But suppose instead that a random transition occurs for particle 1 between time t and time $t + dt$. In that case, the result of the experiment—the perfect anticorrelation—cannot be explained in terms of the initial states of the particles, but must rather invoke some form of nonlocal connection between the particles, so that the transition in the state of particle 1 is sufficient to guarantee that particle 2 makes the corresponding transition. In fact, even without the measurements at time t_m (which are added here only to emphasize the point), some form of nonlocality must be involved; the particles *always* obey the quantum-mechanical joint probabilities, so that when one particle makes a transition between t and $t + dt$, the other particle *must*, at the same time, make the corresponding transition.[14]

Further difficulties would arise in the consideration of other frames of reference, but all such difficulties are resolved, in the present case, by disallowing any random transitions between t_0 and t_m. (None of this discussion is to suggest that modal interpretations can, by an appropriate choice of dynamics, be made completely "local" in any strong sense of the word; no consideration has been given here to Bell-type theorems. Nonetheless, a desire for some weak form of locality might well motivate one to choose a dynamics in which no transition occurs between t_0 and t_m.)

However, even if we accept the condition of no random transitions between t_0 and t_m, we have not nailed down a dynamics, even for this simplified scenario. For example, if both particles are determined from time t_0 to make a transition between t and $t + dt$, then the theory remains local for this scenario. And, of course, we have not come close to nailing down a completely general dynamics, that is, one that applies to any physical situation.

A short digression on Lorentz-invariance is in order here. I have certainly *not* considered the question whether the criteria of Lorentz-invariance may select a unique dynamics. Instead, I have considered, briefly, some other requirements that the theory of relativity might be thought to impose. But clearly Lorentz-invariance is the most important requirement (and perhaps the only one) that the theory of relativity imposes on an interpretation of quantum mechanics. Can this requirement select a unique dynamics? I have no conclusive argument that it cannot, but preliminary investigation by Dickson and Clifton (1998) does not bode well for this hope. It seems to suggest that how one selects A_W is far more important than the details of a dynamics for Lorentz-invariance. Indeed, their analysis shows that the interpretation of Dieks and Vermaas (and therefore the proposal of Bacciagaluppi and Dickson as well) cannot be Lorentz-invariant. Bub's interpretation can apparently be made Lorentz-invariant (or at least it does not fall to the argument of Dickson and Clifton), but only under certain (rather unsatisfactory) choices for the preferred observable, and in these cases there are in fact many possible dynamics.

This discussion illustrates the first of two primary contentions in this section, namely, that reasonable nonempirical principles will rarely, perhaps never, entirely remove the plurality of dynamics. The second contention is that there is no uniquely reasonable set of nonempirical principles that will guide that choice of a dynamics.

To illustrate this second contention,[15] suppose that some system has the wavefunction

$$c_1|\alpha_1\rangle\;|\beta_1\rangle + c_2|\alpha_2\rangle\;|\beta_2\rangle + c_3|\alpha_3\rangle\;|\beta_3\rangle \tag{12}$$

at some initial time, and suppose that the state at the later time is

$$c_1|\alpha_2\rangle\ |\beta_1\rangle + c_2|\alpha_3\rangle\ |\beta_2\rangle + c_3|\alpha_1\rangle\ |\beta_3\rangle \tag{13}$$

Consider what the interpretation of Dieks and Vermaas (1995) would say in this case. The definite properties for the α-system at both the earlier and the later times are (represented by) the projections, $P_{|\alpha_1\rangle}$, $P_{|\alpha_2\rangle}$, $P_{|\alpha_3\rangle}$.

Now consider two conditions that one might impose on a dynamics. The first is a principle of economy, which says that if the probability of a property, $P_{|i\rangle}$, is increasing or remaining constant during some interval of time, while the probability of $P_{|j\rangle}$ is decreasing or remaining constant over the same interval of time, then there is no transition (during that interval) from $P_{|i\rangle}$ to $P_{|j\rangle}$. This principle is a principle of economy because allowing transitions in these cases would require *extra* counterbalancing transitions from $P_{|j\rangle}$ to $P_{|i\rangle}$ to maintain the correct marginals at time $t + dt$. The same net effect could be achieved without these extra transitions by eliminating the transitions from $P_{|i\rangle}$ to $P_{|j\rangle}$ as the principle requires.

But now consider an alternative principle, guided by the intuition that the Hamiltonian should play a primary role in the dynamics. The evolution from (12) to (13) is most naturally seen as driving the property $P_{|\alpha_1\rangle}$ to the property $P_{|\alpha_2\rangle}$, $P_{|\alpha_2\rangle}$ to $P_{|\alpha_3\rangle}$, and $P_{|\alpha_3\rangle}$ to $P_{|\alpha_1\rangle}$. Therefore, one might want to impose the condition that the dynamics should follow this evolution. That is, a system beginning with the property $P_{|\alpha_1\rangle}$ should later possess $P_{|\alpha_2\rangle}$, then $P_{|\alpha_3\rangle}$, then $P_{|\alpha_1\rangle}$, and so on.

Neither of these principles is *empirically* required, and both of them are quite reasonable nonempirical principles, but they are incompatible; no dynamics can satisfy both. To see why, consider a case where $|c_1|^2 = \frac{1}{6}$, $|c_2|^2 = \frac{1}{3}$, and $|c_3|^2 = \frac{1}{2}$. Suppose that the system begins by possessing $P_{|\alpha_1\rangle}$. Then the Hamiltonian rotates the states, so that the new probabilities are $\frac{1}{2}$, $\frac{1}{6}$, and $\frac{1}{3}$ for $P_{|\alpha_1\rangle}$, $P_{|\alpha_2\rangle}$, and $P_{|\alpha_2\rangle}$, respectively. According to the principle of economy, the system *cannot* follow the Hamiltonian evolution, because doing so would require a transition from $P_{|\alpha_1\rangle}$ to $P_{|\alpha_2\rangle}$; but the probability of $P_{|\alpha_1\rangle}$ rose while the probability of $P_{|\alpha_2\rangle}$ fell. We are left with two apparently reasonable, but incompatible, nonempirical conditions.

My second contention in this section is that we are very likely to be always in this situation, that very likely there exists no uniquely reasonable set of nonempirical principles to guide the choice of a dynamics. This contention, together with the first (that reasonable nonempirical principles will rarely or never entirely remove the plurality of dynamics), suggests that the plurality of dynamics is indeed generic and permanent.

Of course, this claim can only be a contention supported by examples. It may be that somewhere in the basement of AT&T labs, somebody has already eliminated all reasonable dynamics save one, either experimentally (which would require violating standard quantum mechanics) or through a deviously clever argument. I can only say that I doubt it.

4. Let a Thousand Flowers Bloom

4.1. Hope and Despair

We are left with an apparently genuine plurality of dynamics. If the suggestions of section 3 are correct, then this plurality will not be uniquely dissolved by any nonempirical principle or set of principles. There are two attitudes one might take to this plurality. First, the attitude of despair: abandon the project of finding a dynamics as worthless, hopeless, or unnecessary. Second, the attitude of hope: welcome the plurality as a sign of scientific fertility; endorse all attempts to find a dynamics based on reasonable principles, while nonetheless believing that none will be the unique dynamics. There is a fine line between these attitudes, but the difference makes a lot of difference. Here are two ways to contrast them.

According to the attitude of despair, the plurality of dynamics teaches us that the choice of a dynamics is at best a mere calculational convenience (although in practice it is usually a calculational inconvenience). To base any inference on the choice of a particular dynamics immediately renders the result of that inference valueless; such inferences are like inferences based on the choice of one of a number of empirically equivalent classical gauge potentials. On the other hand, according to the attitude of hope, the results of inferences based on a particular choice of dynamics may have value, in the context of a particular set of principles used to guide the choice of dynamics at hand.

According to the attitude of despair, physics requires uniqueness, or at least the promise of uniqueness. (It is not the attitude that only one theory should be entertained at a time, but rather the attitude that the ultimate—even if never achieved—aim of physics is a unique and complete theory.) If a particular type of physical inquiry turns out to be incapable of yielding a unique theory, then that type of inquiry is relabeled "nonphysical." Correspondingly, if a particular type of inquiry is apparently, or very likely, not capable of yielding a unique theory, then that type of inquiry is labeled "apparently, or very likely, nonphysical," and it is left to the oddballs and dreamers to pursue. On the other hand, according to the attitude of hope, physics does not seek a unique theory. Instead, understanding of the physical world is *best* achieved by the consideration of many, even incompatible, theories.

4.2. The Value of Hope

Many physics-minded readers are likely to be wary of the attitude of hope. What *scientific* value does its pluralistic spirit have? Why is it not simply contradictory, or even worse, relativist?

Although relativists (who often seem to enjoy contradictions as well) might adopt the attitude of hope, it is not their property only. If the purpose of physics, or pure science in general, is to *explain*, then anybody who believes that explanation is necessarily relativized to a *context* may adopt the attitude of hope.

It is by now commonplace (though not uncontroversial) to suppose that requests for explanation are always made in some context and that therefore explanation itself is relative to a context. For example, van Fraassen (1980, chapter 5) has suggested that explanations are answers to why-questions, and that why-questions consist of four elements: the fact (the event to be explained), the foils (some non-occurrent events), a relevance relation (the relation of the explanans to the explanandum), and background knowledge. A change in any of these can change the question, and therefore the correct answer to the question. (See Lipton [1991] for further discussion of this conception of explanation.)

For example, suppose the why-question is, Why did it rain ten inches today (rather than rain one inch or two inches)? In this case, the foils are "raining one inch" and "raining two inches," and an adequate answer to the question might cite the amount of moisture in the clouds. But now consider the question, Why did it rain ten inches today (rather than snow ten inches)? This question requires a different sort of answer—presumably one that makes reference to the temperature—because it is asked in the context of a different set of foils.

There is certainly no reason to suppose that the answers to these two questions might not be incorporated into a single theory. The point so far is just that requests for explanation can be made in very different contexts, and these contexts may, but perhaps need not, be compatible; that is, they may, but perhaps need not, be susceptible of being treated within a single theory. My further contention is twofold: first, that there are many, incompatible, contexts of explanation ("incompatible" meaning that a single theory cannot provide satisfactory explanations in all contexts), and second, that the legitimacy of any given context cannot be entirely adjudicated within physics, so that there can be no criterion within physics to decide what sorts of context, or question, are legitimate, or interesting, or important for understanding the physical world. My claim is not that physics is irrelevant to such issues, but that it cannot, by itself, resolve them.

A further example, involving the dynamical schemes for modal interpretations, will perhaps be useful. Consider a simple (and idealized) repeated experiment: measure the spin in the z-direction of a single particle, then measure it again. According to standard quantum mechanics, we will always see on the apparatus the same result both times—suppose it is $|z,+\rangle$. Somebody who saw this experiment might well ask, Why did it come up "plus" both times? For modal interpretations, the answer will depend on what type of dynamics one chooses.

Consider a dynamics that obeys the principle of economy of section 3.2. In that case, the particle will make no transitions between the first and second measurements, and a reasonable explanation could cite this fact about the particle. (A deeper explanation might cite the principle of economy itself.)

But now consider the completely uncorrelated dynamics (mentioned in section 2.2), in which the state of a system at one time is uncorrelated with its state at any other time. In this case, an explanation would have to invoke some assumptions about function of memory, the result of which is something like the following: the joint probability (at any given time) of recalling having seen the result "minus" for the first measurement while actually witnessing the result "plus" for the second measurement is zero for a well-functioning brain. (A deeper explanation would perhaps involve the physics of the brain.)

Each of these explanations may be valuable in some way. Each is apparently scientific by usual standards, and each may be the right sort of explanation to give in certain contexts (that is, in reply to certain why-questions, involving different foils, relevance relations, or background knowledge). Whether the phenomena cited in each of these explanations can be treated within a single theory is, at least for now, not clear.

This last point is important. It is *not* my claim that the incompatibility among various dynamical schemes for modal interpretations maps one-to-one onto the incompatibility that I suppose there to be among some explanatory contexts. Rather, my claim is that the adoption of any given dynamical scheme will lead one to cite certain phenomena in one's explanations (and therefore lead one to pursue certain lines of scientific inquiry), and that the adoption of some other dynamical scheme may lead one to cite different phenomena (and to pursue different lines of inquiry). In general, different dynamical schemes give rise to different sorts of explanation, that is, answer different sorts of question.

That claim is already enough to make my most important point, namely, that a plurality of dynamical schemes may be scientifically useful, rather than harmful. However, I have made a stronger claim, namely, that the explanations offered by different dynamical schemes may (sometimes) be strongly incompatible, that is, incapable of being incorporated into a single theory. For example, it may be that the "economical" explanation and the neuro-physical explanation of the result of the repeated experiment cannot be incorporated into a single theory. Nonetheless, they can both teach us something about the world.

This last claim is based on the view that scientific theories need not map onto the world in any straightforward way in order for them to teach us something about the world. It may be that no single theory is capable of describing everything there is to describe about the physical world. One may allow that our theories can teach us about the world, while also allowing the possibility that the world cannot be fully described by any single theory. After all, there is apparently no reason, a priori, to expect that our modes of theorizing are guaranteed to be capable of capturing in a single theory all of the subtlety of the physical world.

In spite of this further claim, I wish to emphasize here that a much more prosaic view of scientific theorizing can nonetheless recognize some advantage for the attitude of hope—for taking on *multiple* explanations without asking "which is right?"—in spite of the incompatibility of the underlying dynamical schemes. Is to do so a contradiction? No. It is, rather, to learn as much as possible from whatever sources are available. It is to recognize that what science is after, ultimately, is as many good explanations as can be found, and that at the least, different theories will generate different explanations, or, what is the same thing, answers to different questions.

4.3. Caveats

The attitude of hope is not the attitude that anything goes. The hope in the attitude of hope is a hope that *some* dynamical schemes will turn out to be helpful or important for understanding the physical world. Considerations of practicality, usefulness, or plausibility might rule out some dynamics as not important or revealing contexts from which to study the physical world. Moreover, the difference among some dynamical schemes might be so slight as to make little or no difference in the explanations offered by each, at least most of the time.

It might turn out, therefore, that what appears now to be a vast plurality is in fact only a handful, or less. Indeed, it might turn out that *all* modal interpretations—not to mention quantum mechanics itself—are false, or implausible in the light of new evidence. The discussion here is therefore ultimately not a discussion *about* modal interpretations, but merely occasioned by them. Although what has been said here should not be taken as evidence for, or advocacy of, modal interpretations, it is, I hope, testimony to their fertility.

Notes

Section 2 is the result of joint work with Guido Bacciagaluppi. Some of the ideas in section 4 were motivated by a talk delivered by Arthur Fine at the conference "Quantum Theory without Observers" (Bielefeld, August 1995). Thanks to Rob Clifton for useful comments, especially about the definition of supervenience. Thanks to Guido Bacciagaluppi and Richard Healey for useful comments on an earlier draft. Thanks to audiences at the Minnesota Workshop on the Measurement Problem, Oxford University, Indiana University, and the University of Western Ontario for helpful comments and questions. Finally, thanks to Richard Healey and Geoffrey Hellman for the invitation to contribute to this volume.

1. Among them are some progeny of van Fraassen himself (1973, 1981, 1991, chapter 9). Other examples are the interpretations of Kochen (1985), Dieks (1988, 1989), Healey (1989), Bub (1992, 1994), Bub and Clifton (1995), and Vermaas and Dieks (1995).

2. This apparently innocuous assumption has come under serious scrutiny recently, and rightly so. The objection is that there may not be a once-for-all one-to-one map from projections operators to properties. (The assumption that there is such a map has been dubbed "naive realism about operators" by Sheldon Goldstein, in a talk delivered to the Philosophy of Science Association in 1996.) Instead, the way that a given projection operator represents a physical property—or indeed whether it does at all—may depend on the physical situation. Modal interpretations do not fall to this objection, however, although I will not pursue the point here; see Dickson (1998).

3. This constraint has been much discussed. For recent discussions and references to earlier ones, see Bacciagaluppi and Hemmo (1994, 1995a, 1995b) and Dickson (1994).

4. Bell and Clifton (1995) have shown that in fact the algebras chosen by modal interpretations are realizations of what they call "quasi-Boolean algebras," which are defined by them purely abstractly, without reference to Hilbert space. See also Zimba and Clifton (1998).

5. For some faux-Boolean algebras, certain other types of measure also permit a classical model, and may therefore be called "faux-classical." The simplest case is when the subspace orthogonal to the span of S is one-dimensional. In these cases, nonzero probability could be assigned to this one-dimensional space, while still preserving the possibility of a classical model.

6. A more extensive discussion of some aspects of how to treat compound systems is in Clifton (1996). His argument bears in particular on whether property composition and decomposition are compatible with certain other constraints on the algebra of definite properties.

7. It has been shown that this approach leads to a Kochen-Specker contradiction (Bacciagaluppi 1995). Dieks must therefore adopt some form of contextuality, in which property-ascription is relative to a factorization. That is, a system has properties only "as a member of a certain factorization." (In his chapter in this volume, however, Dieks seems to pursue an approach that adopts a preferred factorization instead. More precisely, he says that we simply choose a factorization that is convenient, or appropriate, for the problem at hand.)

8. Nick Reeder made this point to me in private correspondence.

9. This section is essentially a summary of some parts of joint work with Guido Bacciagaluppi (Bacciagaluppi and Dickson 1998). The guy who makes cappuccino at the Copper Kettle also played an important supporting role in our work.

10. Or so I assume here. Whether this dynamics is really empirically adequate

appears to depend on difficult questions about the nature of human memory, and whether it is described within quantum mechanics.

11. "Effectively," but not precisely. In Bub and Clifton's interpretation, recall that the elements of S are the projections of the state of the universe onto some preferred observable. If this preferred observable is constant in time, then although the projections onto its eigenspaces may change in time (because the state of the universe evolves), there is a simple mapping from a dynamics on a time-independent algebra to the "true" time-dependent one. This mapping takes advantage of the fact that Bub and Clifton's algebra of definite properties will have at most one nonzero atom in each eigenspace of their preferred observable.

12. This fact is nontrivial, and indeed holds only under certain (but fairly general) conditions. See Bacciagaluppi, Donald, and Vermaas (1996) for a lengthy discussion, and Bacciagaluppi and Dickson (1998) for a less detailed discussion.

13. Bell's expression for T_{ji} is not exactly (4), but rather

$$T_{ji} = \begin{cases} J_{ji} \,/\, \mathrm{p}_i & \text{for } J_{ji} > 0 \\ 0 & \text{for } J_{ji} \leq 0. \end{cases}$$

The difference is irrelevant, at least if p_i has only isolated zeros. In fact, the two choices can differ only at points $\mathrm{p}_i = J_{ji} = 0$, where (4) may be infinite, while Bell's expression is zero. The choice (4) is continuous at these exceptional points.

14. This argument might not hold for every modal interpretation. For example, in van Fraassen's approach, the particles might have few if any interesting properties prior to the measurement. Alternatively, in a similar empiricist spirit, one might wish to impose the quantum-mechanical probabilities (including the joint probabilities) only for *observed* properties (though I know of no modal interpretation that does so), in which case transitions in the state of particle 1 need not affect the state of particle 2, prior to any measurement.

15. The following situation was suggested by comments made by David Albert at a session of the Minnesota Workshop on the Measurement Problem in May 1995.

References

Bacciagaluppi, G. 1995. "A Kochen-Specker Theorem in the Modal Interpretation of Quantum Mechanics." *International Journal of Theoretical Physics* 34: 1205–16.

Bacciagaluppi, G., and M. Dickson. 1998. "Dynamics for Density-Operator Interpretations of Quantum Theory," unpublished manuscript.

Bacciagaluppi, G., and M. Hemmo. 1994. "Making Sense of Approximate Decoherence." In D. Hull, M. Forbes, and R. Burian, eds., *Proc. 1994 Biennial Meeting of the Philosophy of Science Association*, vol. 1. East Lansing: Philosophy of Science Association, 345–54.

———, and ———. 1996. "Modal Interpretations, Decoherence, and Measurements." *Studies in History and Philosophy of Modern Physics* 27: 239–77.

Bacciagaluppi, G., M. Donald, and P. Vermaas. 1996. "Continuity and Discontinuity of Possessed Properties in the Modal Interpretation." *Helvetica Physics Acta* 68: 679–704.

Bell, J. 1987. "Beables for Quantum Field Theory." In *Speakable and Unspeakable in Quantum Mechanics*. Cambridge: Cambridge University Press. 173–80.

Bell, J. and R. Clifton. 1995. "QuasiBoolean Algebras and Simultaneously Definite

Properties in Quantum Mechanics." *International Journal of Theoretical Physics* 34: 2409–21.
Bub, J. 1992. "Quantum Mechanics without the Projection Postulate." *Foundations of Physics* 22: 737–54.
———. 1994. "On the Structure of Quantum Proposition Systems." *Foundations of Physics* 24: 1261–79.
Bub, J., and R. Clifton. 1996. "A Uniqueness Theorem for Interpretations of Quantum Mechanics." *Studies in History and Philosophy of Modern Physics* 27: 181–219.
Clifton, R. 1995. "Independently Motivating the Kochen-Dieks Modal Interpretation of Quantum Mechanics." *British Journal for the Philosophy of Science* 46: 33–57.
———. 1996. "The Properties of Modal Interpretations of Quantum Mechanics." *British Journal for the Philosophy of Science* 47: 371–98.
Dickson, M. 1994. "Wavefunction Tails in the Modal Interpretation." In D. Hull, M. Forbes, and R. Burian, eds., *Proc. 1994 Biennial Meeting of the Philosophy of Science Association, vol. 1*. East Lansing: Philosophy of Science Association. 366–76.
———. 1995a. "Is There *Really* No Projection Postulate in the Modal Interpretation?" *British Journal for the Philosophy of Science* 46: 197–218.
———. 1995b. "Faux-Boolean Algebras, Classical Probability, and Determinism." *Foundations of Physics Letters* 8: 231–42.
———. 1995c. "Faux-Boolean Algebras and Classical Models." *Foundations of Physics Letters* 8: 401–15.
———. 1996. "Logical Foundations of Modal Interpretations of Quantum Mechanics." In *Philosophy of Science, PSA96 Supplemental Issue*, 322–29.
———. 1998. *Quantum Chance and Nonlocality: Probability and Nonlocality in the Interpretations of Quantum Mechanics*. Cambridge: Cambridge University Press.
Dickson, M., and R. Clifton. 1998. "Lorentz-Invariance in Modal Interpretations." Forthcoming in *The Modal Interpretation of Quantum Mechanics*, ed. D. Dieks and P. Vermaas. Dordrecht: Kluwer.
Dieks, D. 1988. "The Formalism of Quantum Theory: An Objective Description of Reality?" *Annalen der Physik* 7: 174–90.
———. 1989. "Resolution of the Measurement Problem through Decoherence of the Quantum State." *Physics Letters* A142: 439–46.
Healey, R. 1989. *The Philosophy of Quantum Mechanics: An Interactive Interpretation*. Cambridge: Cambridge University Press.
Kochen, S. 1985. "A New Interpretation of Quantum Mechanics." In P. Lahti and P. Mittelstaedt, eds., *Symposium on the Foundations of Modern Physics*. Singapore: World Scientific, 151–69.
Lipton, P. 1991. "Contrastive Explanation." In D. Knowles, ed., *Explanation and Its Limits*. Cambridge: Cambridge University Press.
Reed, M., and B. Simon. 1975. *Methods of Modern Mathematical Physics*. San Diego, Calif.: Academic Press.
van Fraassen, B. 1972. "A Formal Approach to the Philosophy of Science." In R. Colodny, ed., *Paradigms and Paradoxes: Philosophical Challenges of the Quantum Domain*. Pittsburgh: Pittsburgh University Press, 303–66.
———. 1973. "A Semantic Analysis of Quantum Logic." In C. Hooker, ed., *Contemporary Research in the Foundations and Philosophy of Quantum Theory*. Dordrecht: D. Reidel, 80–113.
———. 1980. *The Scientific Image*. Oxford: Clarendon Press.
———. 1981. "A Modal Interpretation of Quantum Mechanics." In E. Beltrametti and B. van Fraassen, eds., *Current Issues in Quantum Logic*. New York: Plenum, 229–58.

———. 1991. *Quantum Mechanics: An Empiricist View*. Oxford: Clarendon Press.

Vermaas, P. 1996. "Unique Transition Probabilities in the Modal Interpretation." *Studies in the History and Philosophy of Modern Physics* 27: 133–59.

Vermaas, P., and D. Dieks. 1995. "The Modal Interpretation of Quantum Mechanics and Its Generalization to Density Operators." *Foundations of Physics* 25: 145–58.

Vink, J. 1993. "Quantum Mechanics in Terms of Discrete Beables." *Physical Review* A48: 1808–18.

Zimba, J., and R. Clifton. 1998. "Valuations on Functionally Closed Sets of Quantum-Mechanical Observables and Von Neumann's No-Hidden-Variables Theorem." Forthcoming in *The Modal Interpretation of Quantum Mechanics*, ed. D. Dieks and P. Vermaas. Dordrecht: Kluwer.

William G. Unruh

Varieties of Quantum Measurement

The problem of quantum measurement has been with us since the foundations of the theory were laid in the mid 1920s. It has generated much discussion, with little resolution of the questions raised. I will argue in this chapter that this situation has arisen in part because of the confusion brought about by giving two very different concepts the same name, with the expected result that the valid questions related to the two concepts become entangled. It furthermore has led to a restriction on the types of measurements considered within the theory. I am not going to propose any radical or even very new interpretations of the theory of quantum mechanics; rather, I will engage in an ancient philosophical pastime, namely, to propose that we use distinct terms for distinct concepts. I am then going to review some of the insights that have been obtained recently (especially by the group around Aharonov) regarding some novel types of measurement.

1. Measurement, Determination, and Knowledge

The concept of measurement in quantum mechanics has had a long and confused history. Essentially, two separate concepts have been conflated under the same title, concepts with a very different status in the theory a priori. In part the intense confusion surrounding the word results from the attempt to reconcile these two different concepts, or rather to apply the properties of the one concept to the other.

The first concept subsumed under the term measurement is an axiomatic concept. Quantum mechanics, as with all of our theories in physics, is based on a set of mathematical structures. In the case of quantum mechanics, these structures are those of complex Hilbert spaces and operators on those Hilbert spaces. In addition to such mathematical structures, the theory must also make contact with the physical world. Structures in the theory must be correlated by structures in our experience of the world itself. As with all theories, quantum mechanics is a means of answering questions about our experiences of the actual world we live in. The theory requires mapping the mathematical structure onto our experiences. This takes the form both of a general map—of general structures of the world which we expect to have a broad range of validity—and structures that reflect the particulars and peculiarities of our experiences.

In classical physics, the former is called the dynamical theory, while the latter is called the initial conditions. The theory encompasses the identification of dynamic variables and equations of motion, while the initial conditions encompass those aspects of our experience which are felt to be peculiar to the individual time and place of those experiences.

Quantum mechanics contains both of these aspects as well, but in a very different form from that of classical physics. The dynamics is represented by the operators, while, in the simplest case, the particulars of the situation are represented by the vector in the Hilbert space, the wave-function. I will denote these particulars by the term "knowledge" or "conditions," rather than the term "initial conditions," since, as we will see, conditions need not be initial nor are they in general equivalent to initial conditions (as they are in classical physics).

In addition to explanations, the theory must produce answers, must give us the answers to questions that we may have about the physical situations that we are interested in. Here the theory actually makes contact with the physical world. In quantum mechanics these answers are in terms of probabilities. The usual phraseology goes something like, "When one measures a quantity, and the system is in the state $|\phi>$, the outcome of that measurement is one of the eigenvalues, say a, of the operator, say $\mathbf{A}$, representing the physical variable measured, and the probability is given by the usual expression $|<a|\phi>|$."

The word "measure," however, brings with it the image of a physical process. Measurements are performed by means of measuring apparatuses. As aspects of the physical world, such measuring apparatuses should themselves be describable by quantum mechanics. But it is difficult to have a system in which at the same time a concept is an axiomatic feature of the theory *and* one describable by the theory. I would therefore suggest that the word "determine" be used instead for this axiomatic feature of the theory. Thus I would rephrase the above sentence as "When one determines a quantity, and the knowledge (or conditions) under which one wishes to determine that quantity are represented by the vector $|\phi>$, then the determination of a quantity represented by $\mathbf{A}$ gives one of eigenvalues of $\mathbf{A}$, say a with probability $|<a|\phi>|^2$."

Determination, in this axiomatic sense, says nothing about how the determination was made. It is simply a statement of a mapping from the theory to our experience, in which some knowledge sets the conditions on the questions we wish to ask, and some knowledge represents the answers to the questions we want to ask.

What then is a measurement? I will reserve the term measurement for a physical process, a process describable in terms of quantum theory itself. A measurement is a process in which one has two separate physical systems, represented by two separate sets of dynamical operators. Furthermore the dynamical evolution is such that, given certain conditions on the measuring apparatus, a determination of some quantity associated with the measuring apparatus will give information about the system of interest.

Von Neumann (1955) showed that under certain conditions, a measurement on a system could be treated as a determination of that system. That is to say, certain types of measurement (in which one makes a determination of some aspect of the measuring apparatus only) acted in all ways as though one had instead made a determination of the system itself. There is a consistency in quantum mechanics, such that the axiomatic concept I call determination is closely related to the physical process I call measurement. Notice, however, that in von Neumann's analysis, one has not done away with the concept of determination. One still must apply the axiomatic concept of determination to the measuring apparatus before one can draw any conclusions at all from the theory. It is just that such measurements allow us to reduce a complicated system (apparatus plus system of interest) to a simple system (the system of interest alone) under certain conditions. This mapping of a complex system onto a simpler system does not, however, in any way change the requirement for the axiomatic concept of "determination." It simply changes the system to which we must apply the concept.

At least in part the measurement problem in quantum mechanics is the disquiet that physicists feel for the concept of "determination." It feels like an extra and extraneous concept, a nonphysical concept. In classical physics, one can imagine that the theory and reality are in complete correspondence. The position of a particle really is a number, and our experience of that position is simply the experience of that number. The physical map from experience to theory is just an identification of those numbers in the theory with the experience. (That some fairly sophisticated manipulations of experience are necessary to extract that number is a technical detail.) In quantum mechanics, on the other hand, there seems to be no direct map from our experience to the theory. The operators themselves have far too much structure for experience. The state, or Hilbert space vector itself, has the wrong properties to map onto our experience. The only map is the rather indirect and seemingly unnatural one of "determination." One would like either to subsume determination under some physical concept of the theory (with the consequence that the only relation linking experience and the theory would be lost) or to introduce some other relation between the theory and experience from which one could derive determination in a natural way. That neither of these objectives has ever been achieved is a large part of the so-called problem of measurement in quantum mechanics.

I do not want to spend any more time on this issue. Rather, I want to point out that the concern about this problem has warped our thinking about quantum mechanics and about the types of measurement possible in the theory. Because the von Neumann type of measurement creates the possibility of reduction of a complex system to a simpler system, the idea has become implanted that all measurement must be of the same sort. The thinking seems to be that because determination has a certain form, measurement must have the same form. It is becoming clear, however, especially through the work of the group around Aharonov, that this is too restrictive.

Measurement is a physical process by which one has two systems interacting, and by making a determination on the one system, one can obtain information about the other system. In certain cases, the information obtained is the same as a determination, but in other cases it can differ significantly. Furthermore, because of the similarity of wave mechanics to classical wave theory, the impression has also arisen that conditions in quantum mechanics are entirely equivalent to conditions in classical mechanics, namely, initial conditions.

2. Initial and Final Conditions

It has long been known to some (but ignored or resisted by most) that the conditions in quantum mechanics differ significantly from those of classical physics (Aharonov, Bergmann, and Lebowitz 1964). In classical physics, all conditions can, by use of the equations of motion, be mapped onto initial conditions. Whether one measures the position now and the momentum two days hence, or measures them both now is irrelevant. For any condition, imposed at any time, one can always, by use of the equations of motion, produce initial conditions that are entirely equivalent in all of their predictions to those general conditions. However, as Aharonov, Bergmann, and Lebowitz (1964) already showed more than thirty years ago (and as has been independently rediscovered often since [see Unruh (1986)]), setting conditions at different times may not be equivalent to any initial conditions. The simplest example is that of a spin-$\frac{1}{2}$ particle whose x component of spin is known at 9 A.M. and y component at 11 A.M. Say both are known to have value $+\frac{1}{2}$. The probability that if one determines the component $\cos(\theta)S_x + \sin(\theta)S_y$ at 10 A.M., the answer will be $+\frac{1}{2}$ is

$$P_{S\theta = \frac{1}{2}} = \frac{(1 + \cos(\theta))(1 + \sin(\theta))}{2(1 + \sin(\theta)\cos(\theta))}. \tag{1}$$

Note there exists no initial condition—wave-function or density matrix—that would give this answer. It is unity for both $\theta = 0$ and $\theta = \pi/2$. The conclusion drawn from this simple example is true in general: conditions in quantum mechanics are not equivalent to initial conditions.

Already in 1986, Aharonov (Aharonov et al. 1986) mentioned a surprising new effect that combines the inequivalence of conditions to initial conditions together with what he calls "weak" measurements. If we set both initial and final conditions, and at the intermediate time perform a particular type of inexact measurement of a quantity, the outcome of that measurement can be very counterintuitive. Although the measuring apparatus and the interaction are designed so that if the initial state is an eigenstate of the measured quantity, the outcome will be approximately given by that value for the measured quantity, in this pre- and post-conditioned experiment, the expected value for the measurement is impossible according to all the usual tenets of quantum mechanics.

Let me make this clear by an example. Our measuring apparatus is a trivial infinite mass free particle. It is coupled to a spin s particle (in my example $s = 20$). The coupling is of the form

$$H_I = \frac{\epsilon}{\sqrt{(2)}}(S_x + S_y)p\delta(t - t_0). \tag{2}$$

In other words, the interaction is such that if the initial state of the free particle is $\psi(x)$, and the state of the spin is in an eigenstate of the operator $S_{//}$ say with eigenvalue σ, then the final state of the free particle is $\psi(x - \epsilon\sigma)$. Thus by measuring the displacement of the free particle due to the interaction one can estimate σ and thus measure $S_{//}$. If the particle begins in an eigenstate (or almost an eigenstate) of X (that is, ψ is sharply peaked about some value x_0 with an uncertainty much less than ϵ), then the displacement during the interaction can be measured precisely by determining the value x of X after the interaction, and $\sigma = (x - x_0)/\epsilon$ will be a measurement of S. On the other hand, if the initial ψ has a spread of Δx, then the final determination of X will give the displacement only to $\pm\Delta x$, and we will have $\sigma = (x - x_0)/\epsilon \pm \Delta x/\epsilon$. This is the sense in which the measuring apparatus is inexact. The determination of some variable of the measuring apparatus only gives an inexact estimate of the value of some dynamic variable of the system.

Now consider the following situation. Set conditions such that before the interaction with the measuring apparatus, the value of S_x is known to

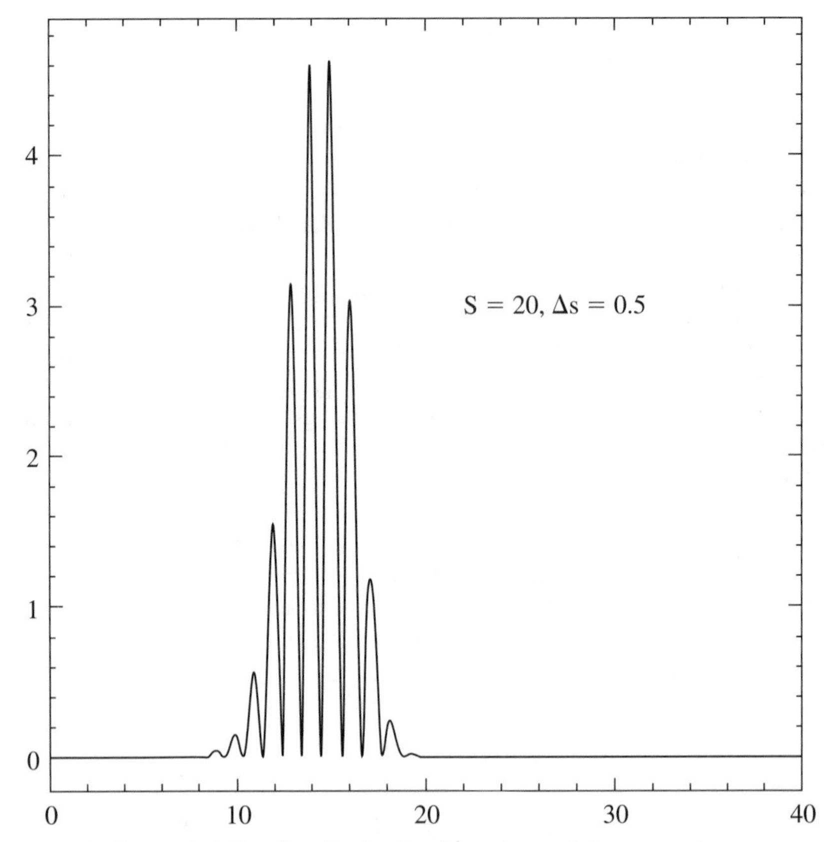

Figure 1. The probability distribution for the pointer of the measuring apparatus with pre- and post-conditions with a small error (≈ .5) for the measurement of the spin. The maximum value of the spin is 20.

be its maximum possible value, s. Furthermore, after the interaction, the value of the y component, S_y, is known to be the maximum possible value, s. What is the distribution of possible outcomes for the measuring apparatus? One would expect this to be a convolution of a probability distribution over the various possible values for $S_{//}$ with the initial probability distribution for the position of the free particle, i.e., one would expect something like $\Sigma_\sigma P_\sigma |\psi(x - \epsilon\sigma)|^2$ where P_σ is a probability for the spin to have value $-s \leq \sigma \leq s$. In particular, the average value (expectation value) for X should lie somewhere between $x_0 - \epsilon s$ and $x_0 + \epsilon s$. If the measurement is sufficiently accurate, this expectation is fulfilled. Figure 1 plots the probability distribution for the location of the particle ($x_0 = 0$

and $\epsilon = 1$) in the case where the initial spread of the wave-function for the particle is small. Figure 2 is the plot of the distribution for the value of the position of the particle in the case where the initial spread for X is large (of order $\pm\epsilon\sqrt{(s)}$). Note that the center of the probability distribution is at $x \approx 28$, and the probability that x would lie between -20 and 20, the naively expected values, is very small. Using the determined value of X to infer the value of $S_{//}$ gives a value at all times larger than the maximum eigenvalue of $S_{//}$.

If we regard S as a classical vector spin, and we know that S_x and S_y both have value s, then $S_{//}$ will have value $\sqrt{S_x^2 + S_y^2} = \sqrt{2}s \approx 28$. Note that

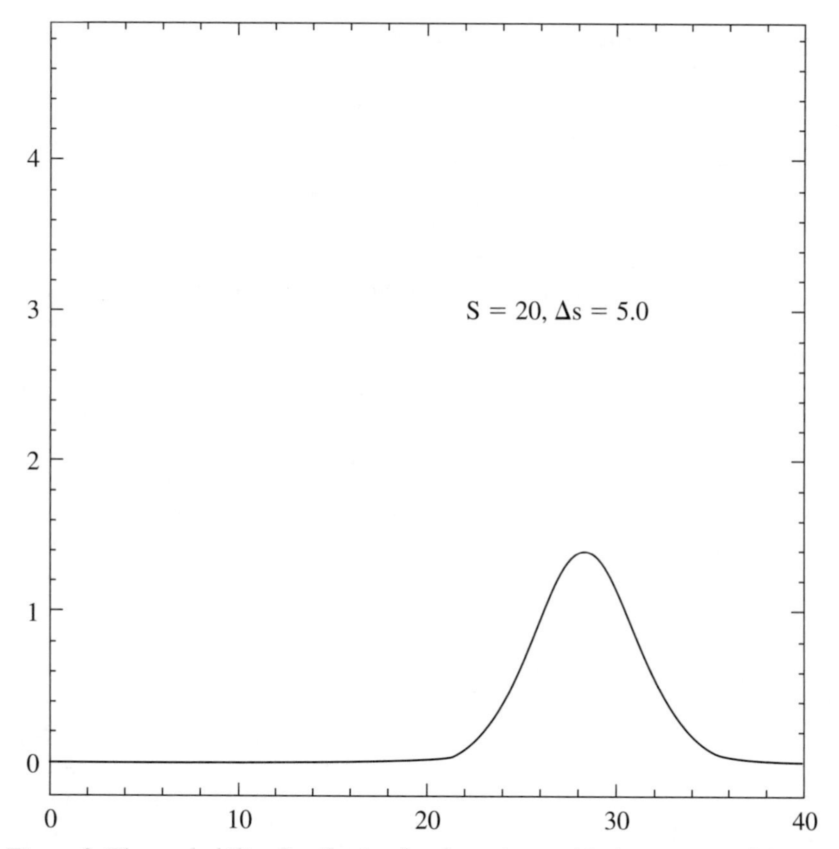

Figure 2. The probability distribution for the pointer with the same conditions as in Figure 1 but with a large error (5) for the inferred value of the spin. Note that the distribution centers around the value of 28 and has only a very small probability of lying between 20 and −20.

this works in this way only if the initial state $\psi(x)$ is sufficiently smooth. (In my case I have chosen it to be a Gaussian.) In particular, sharp features in ψ will destroy this property.

One reaction to this example is that it is not a real measurement. However, it meets all of the criteria of a measuring apparatus, in that if the state of the spin is an eigenstate, the measuring apparatus produces the value to the accuracy to which the apparatus is designed. What we have here is a strange result which arises from the combination of an inexact measuring apparatus, combined with the inequivalence of conditions in quantum mechanics to be equivalent to initial conditions. (For any initial conditions, the expectations that the result would simply have been the sum of the probabilities of the result for the eigenstates would have been true.) This is a measurement situation in which the measurement is not equivalent to any determination on the spin system itself.

Some argue that such situations, in which one sets both initial and final conditions, are unnatural and do not arise in physically interesting situations. I would like to mention one well-known and very old question which falls right into this category of experiment, the question of the time taken by a particle to tunnel through a barrier (Landauer and Martin 1994). It is of course clear that this is exactly the type of situation envisaged in the above description of measurements. One sets up both an initial condition—namely, that the particle is approaching the potential barrier with some energy or in some specific state—and a final condition that the particle must actually penetrate the barrier and come out the other side. The overwhelming probability is that the particle will instead reflect from the barrier.

Deciding what one means by the time of barrier penetration is of course the difficulty. One can imagine setting up an initial wave packet and timing when the wave packet emerges from the barrier. Any such wave packet will consist of a wide range of energies, however. The particles that emerge from the barrier will consist largely of those particles whose energy was sufficiently high to simply go over the barrier. One could use a broad initial wave packet, and regard the peak of the packet emerging from the barrier or the mean position in the packet as defining the time of transit. For a free particle with mass, the dispersion in free flight makes this technique difficult to make unambiguous. (Furthermore, the peak of a packet is a terrible measure of the time, since it would result in the time of travel through a chopper which allowed only the leading edge of the packet to get through to be much faster than light.)

Following the previous example of the spin system, one can instead set up a measurement process of the time of transition in the time during the barrier penetration. Consider a clock attached to the particle, and consider the clock designed so that it is operational only while the particle is in the vicinity of the barrier. For example, choose a Hamiltonian for the system to look like

$$H = \frac{p^2}{2m} + V(x) + \sigma(x)H_c \tag{3}$$

where $\sigma(x)$ is nonzero and of unit value only in a region including the region where $V(x)$ is nonzero. H_c is the Hamiltonian for the clock. For definiteness, I will assume that σ is unity in the interval from 0 to L, while $V(x)$ is nonzero only in a region around $L/2$.

Let me first choose $V(x)$ to be zero everywhere. I will show that the clock now measures the time that the particle takes to traverse the region $0 < x < L$ as one would expect. Assume that the energy eigenstates of the clock Hamiltonian are given by $\phi_\epsilon(q)$ where q represents the internal degrees of freedom of the clock. We can now write the solutions for the energy eigenstates for the whole system as

$$\Psi_E(t, x, q) = e^{-iEt} e^{i\int \kappa(x)dx} \phi_\epsilon(q) \tag{4}$$

where $\kappa(x) = \sqrt{2m(E - \sigma(x)\epsilon)}$. (I have here neglected the reflection from the step discontinuities in σ, as these do not play any important role in the analysis.) In particular, if we concentrate on the region where $x > L$, the energy eigenstates have the form

$$\Psi(t, x, q) = e^{-iEt} e^{ikx - \sqrt{2m(E-\epsilon)}L} \phi_\epsilon(q). \tag{5}$$

For the clock to act as a good clock, one clearly does not want to place it into an energy eigenstate, where none of the dynamical variables, especially those representing the "hands" of the clock, change in time. One must place the clock into a superposition of energy states. In particular, if q represents the position of the hands of the clock, then we want $e^{-iH_c t}\phi(q) \approx \phi(q - t)$. As time goes on, under the Hamiltonian evolution of the clock the hands should advance by an amount equal to the time that has passed. Furthermore, one wants $\phi(q)$ such that a measurement at the time t of q to give an answer without too large an error in the value of q so that one can use that measured value of q to infer what the elapsed time actually was. One wants $\delta q^2 = \int |\phi(q)|^2(q^2 - \bar{q}^2)dq$ to be reasonably small, since the uncertainty in the time inferred from the measured value of q is directly proportional to δq.

$\phi(q)$ can be written as

$$\phi(q) = \int \alpha_\epsilon \phi_\epsilon(q). \tag{6}$$

Let us assume that the spread in the ϵ values contained in ϕ is not too great and centered around ϵ_0. Then we can write

$$\sqrt{2m(E - \epsilon)} \approx \sqrt{2m(E - \epsilon_0)} - \sqrt{\frac{2m}{E - \epsilon_0}}(\epsilon - \epsilon_0). \qquad (7)$$

However, as long as $\delta\epsilon = \epsilon - \epsilon_0$ is small compared with $E - \epsilon_0$, $\sqrt{2m/E - \epsilon_0}$ is just the inverse velocity of the particle in the region where σ is nonzero. Thus the wave-function in the region of large x can be written as

$$\Phi(t, x, q) = e^{iEt}e^{ikx}e^{i(K_0 - k)L}e^{i\epsilon_0 L/v} \int \alpha_\epsilon e^{-i\epsilon L/v}\phi_\epsilon(q)d\epsilon. \qquad (8)$$

But, we have

$$\int \alpha_\epsilon e^{-i\epsilon L/v}\phi_\epsilon(q)d\epsilon = e^{-iH_c L/v}\phi(q) \approx \phi(q - L/v). \qquad (9)$$

The state of the system is such that the clock hands have advanced by an amount proportional to L/v, which is just the time one would argue that the particle took to traverse the distance L. This is true even if the particle itself is in an energy eigenstate (it has no well-defined position at any time), and is true if the particle is in a superposition of energy eigenstates, as long as the spread in the values of E does not introduce too large an uncertainty into v.

Note some important points here. In order that v be fairly definite, ϵ must not have too wide a spread around ϵ_0. In order that the clock give a relatively definite time, it is important that δq not be too large. However, H_c and Q do not commute (since Q is supposed to change as time goes on if the clock is to be a good clock). Thus it is important that the measurement of the time be inherently inaccurate if the clock is to measure the time of transit accurately. That is, the clock must, as in the spin example above, be an inexact measurer of time if the clock is to give a good measure of the transit time.

We can now analyze the same situation if we introduce the barrier $V(x)$. I will assume that the barrier is sufficiently slowly varying that the WKB approximation can be used. In this case the energy eigenstates will have the form in the region where $x >> L$ of

$$\Psi(t, x, q) \approx e^{-iEt}e^{\int \kappa(x)dx}\phi_\epsilon(q) \qquad (10)$$

where

$$\kappa(x) = \sqrt{2m(E - V(x) - \epsilon)}. \qquad (11)$$

In the region under the barrier, κ will be imaginary, and we must choose the sign so that the mode is damped under the barrier. Again expanding around ϵ_0 we have

$$\Psi(t, x, q) \approx e^{-iEt} e^{i\int \kappa_0(x)dx} \int \alpha_\epsilon e^{-i\delta\epsilon \, 1/v(x)\, dx} \, \phi_\epsilon(q) d\epsilon \tag{12}$$

$$\approx e^{-iEt} e^{i\int \kappa_0(x)dx} e^{i\epsilon_0 \, 1/v(x)\, dx} e^{H_c \int dx/v(x)} \phi(q) \tag{13}$$

where

$$v(x) = \sqrt{\frac{2m}{E - V(x) - \epsilon_0}} \tag{14}$$

is the classical velocity and is imaginary under the barrier. Thus the contribution to the change in the state of the clock while the particle is under the barrier is given by

$$e^{-H_c \int dx/\tilde{v}(x)} \phi(q) \tag{15}$$

where $\tilde{v}(x)$ is the Euclidean velocity of the particle under the barrier. Regarding the operator $e^{-iH_c t}$ as the time translation operator on the clock for a time t, we see that while the particle is under the barrier, the best description is that the clock has advanced by an imaginary time, equal to the usual Euclidean tunneling time.

This does not mean that somehow the hands of the clock now point in some imaginary direction. Rather the "evolution operator" $e^{-H_c \tau}$ has the effect of pushing the system down toward its ground state. Because in the ground state the position of the hands of the clock become completely uncertain, the effect of this evolution through an imaginary time is to make the clock reading more and more uncertain: the tunneling through the barrier increases the spread in the clock reading rather than giving a definite reading. As a shorthand, we can ascribe this increased uncertainty to the clock's having measured an imaginary time.

This example has similarities to the spin system. In that case the best reading for the intermediate value of the spin is a value larger than any of the eigenvalues of the operator being measured. In this case the best value for the time of penetration is an imaginary value, again a value not in the spectrum of the operator (the clock hand position) being measured. This is another illustration of the strange—yet on further thought not ridiculous—results one can obtain when one makes measurements in situations in which both initial and final conditions are imposed.

3. Adiabatic Measurements

Another measurement situation leads to results in conflict with the von Neumann equivalence of measurements and determination. This is a situation I call adiabatic measurement. It arises out of another situation noted by the group around Aharonov (Aharonov, Anandan, and Vaidman 1993). (They call it "protected" measurements, a term I feel to be highly misleading. They furthermore use it to argue that the wave-function is "real" in some sense, a conclusion I also have great difficulty with [Unruh 1994].) This is a situation in which the measuring apparatus is coupled to the system sufficiently weakly, and the system's evolution during the interaction with the measuring apparatus is dominated by a Hamiltonian with sufficiently widely spaced energy levels that the interaction with the apparatus can be treated throughout as an adiabatic perturbation.

Consider a system whose Hamiltonian during the course of the interaction is given by H_0. Consider couplings to a set of measuring apparatuses (which for simplicity we will take as free infinitely massive particles again, although nothing changes if we use more complex measuring apparatuses):

$$H = H_0 + \sum_i \epsilon_i(t) A_i P_i \tag{16}$$

where the A_i are a variety of operators associated with the system (in general noncommuting) and the p_i are the momenta of a set of free infinitely massive particles. We can solve this, assuming that the measuring apparatuses are in the momentum eigenstates $|p_i>$, to obtain the adiabatic approximation to the Schrödinger equation for the system

$$|\psi(t)>= \sum_E a_E(t)|E> \tag{17}$$

$$i\dot{a}_E = E(t)a_E(t) - i\sum_{E'} a_{E'} < E(t)\|E'(\dot{t}) >\approx E(t)a_E \tag{18}$$

where

$$E(t) \approx E_0 + \sum_i \epsilon(t) < E_0|A_i|E_0 > p_i \tag{19}$$

where $|E_0>$ are the eigenstates of H_0 and E_0 their eigenvalue. Thus the equation of motion for the state of the system plus measuring apparatus can be written as

$$|\Psi(t)>= \sum_E a_E(0)\, e^{i\int^2(E_0+\Sigma_i\epsilon(t)<E_0|A_i|E_0>P_i dt)}|E(t)> \prod_i |\phi_i> \tag{20}$$

$$= \sum_E e^{iE_0t}|E(t)> \prod_i e^{i\int\epsilon(t)dt<E_0|A_i|E_0>P_i}\phi_i(x_i) \tag{21}$$

$$= \sum_E e^{iE_0t}|E(t)> \prod_i \phi_i(x_i - \int \epsilon_i(t)dt < E_0|A_i|E_0 >). \tag{22}$$

After the interaction with the apparatus is finished, the state is

$$|\Psi(t)>= \sum_{E_0} a_E(0)e^{iE_0t}|E_0> \prod_i |\phi(x_i - \int \epsilon_i dt < E_0|A_i|E_0>)>. \quad (23)$$

Each of the measuring apparatuses has been displaced by an amount $\int \epsilon_i(t)dt < E_0|A_i|E_0 >$, that is, by an amount proportional to the expectation value of the measured operator A_i in the state $|E_0>$. Now, if we assume that the states $\phi(x_i)$ are sufficiently narrow that there is at least one A_i such that $\int \epsilon_i(t)dt\ (< E_0|A_i|E_0 > - < E_0'|A_i|E_0' >$ is larger than the initial uncertainty in $\phi_i(x)$, then the various energy eigenvalues will decohere. The measuring apparatuses will point to a value $< E_0|A_i|E_0 >$, an expectation value, for some value of E_0, with the probability of that E_0 given by $|a_i(0)|^2 = |< E_0|\psi >|^2$.

There are a number of strange features of this result. In the first place, the value to which the measuring apparatus points is not that corresponding to one of the eigenvalues of A_i. The measuring apparatus measures A_i, but the pointer does not give one of A_i's eigenvalues but instead gives an expectation value, $< E_0|A_i|E_0 >$ in any single measurement. Furthermore, if we repeat the experiment, we will, as expected, get a variety of answers to which the pointer points, namely, each of the various expectation values for the various possible values of E_0. Over a large number of trials, we expect to get a number of trials in which we get a specific value $< E_0|A_i|E_0 >$ a number of times given by $N|< E_0|\psi >|^2$ times. Thus the statistical expectation values for the measurements of A_i are

$$< A_i >_{stat} = \sum_{E_0} |< \psi|E_0 >< E_0|A_i|E_0 >< E_0|\psi >. \quad (24)$$

But the quantum-mechanical expectation value of A_i is given by

$$< A_i >_{QM} \sum_{E_0} \sum_{E_0'} |< \psi|E_0 >< E_0|A_i|E_0' >< E_0'|\psi >. \quad (25)$$

In general, only if the vectors E_0 are also eigenvectors of A_i are these two expressions the same. The statistical expectation value of A_i obtained by performing a large number of such adiabatic measurements is *not* the quantum expectation of A_i in the state of the system.

We thus have a situation that violates almost all of the standard lore about measurements. Since the A_i are not necessarily commuting (nothing in the above derivation demands that they commute), we can in a single measurement measure noncommuting variables. Furthermore, if the initial state is an eigenstate of H_0, then every measurement in an ensemble of measurements will give exactly the same value for the measurement of those noncommuting variables, and there will be no statistical uncertainty in the result. The outcome of the measurement is *not* an eigenvalue of the operator corresponding to the measured quantity A_i but rather an expectation value of that quantity. The statistical distribution of the results does not depend on the quantities A_i being measured, but rather on the eigenvectors of the Hamiltonian H_0, which is *not* coupled to any measuring apparatus at all.

It is interesting to note that the standard von Neumann measurement falls into exactly this class as well. In the von Neumann measurement, the interaction with the measuring apparatus is such that the coupling to the apparatus dominates the dynamics during the measurement. To illustrate, the Hamiltonian is of the form

$$H = H_{free} + \epsilon\delta(t)AP. \qquad (26)$$

In this case, the dominant Hamiltonian during the interaction is A, since $\delta(0)$ is infinite. The coupling to the measuring apparatus A clearly commutes with the dominant Hamiltonian A and thus the interaction is adiabatic for an arbitrary time dependence of $\epsilon(t) = \epsilon\delta(t)$. According to our adiabatic analysis, the measurement will give us various expectation values $< E_0|A|E_0 >$ where the E_0 are the eigenvalues of the dominant Hamiltonian A. That is, the E_0 are just the eigenvalues a of A. Thus, the measured quantities will be $< a|A|a >= a$, the eigenvalues of A. The probability of obtaining the value of a in the measurement is $| < E_0|\psi > |^2 = |< a|\psi > |^2$, and the statistical expectation value of A is

$$< A >_{stat} = | < \psi|a > |^2 a =< A >_{QM}. \qquad (27)$$

We thus see that the usual rules on measurement are simply a special case of the results obtained for adiabatic measurements.

Note that the general adiabatic measurement is not equivalent to a determination, which, however, does not make them any less interesting as measurements. In fact, the archetypal quantum measurement example, the Stern Gerlach experiment, used in almost all textbooks as an example of the von Neumann measurement, is actually an adiabatic measurement, in which noncommuting observables, the spin in both of the transverse directions, is adiabatically measured; for details see Unruh (1994).

4. Conclusions

Let us sum up the key points of this chapter:

1. In the standard formulation of quantum mechanics, the term "measurement" is used to denote two distinct concepts. In order to clarify the

problems, I have suggested that it would be useful to use separate terms to denote separate concepts, and have proposed that we use "determination" for the axiomatic concept and reserve "measurement" for the physical notion of using changes induced into one system to deduce properties of another system.

2. I have pointed out the old but little-understood feature of quantum mechanics that the conditions in quantum mechanics are not equivalent to initial conditions. A couple of examples have emphasized this unexpected nature of the results obtained in quantum mechanics when conditions span the time during which one wants to ask questions of the quantum system.

3. I have shown that if we liberate the notion of measurement from determination, the variety of measurements are in fact much larger than simply those that are equivalent to a determination. Although this has been well known for a long time in the case of inexact measurements, the example of adiabatic measurements shows that many of the features of measurements of the von Neumann type are features restricted to that type of measurement alone.

•

I would like to thank Y. Aharonov and L. Vaidman for discussion. I also thank Paul Davies for bringing to my attention the problem of the barrier tunneling time, and inciting me to analyze it from the point of view of initial and final conditions on the system. I would also like to thank the Canadian Institute for Advanced Research for a fellowship and other support during the course of this work. This work was performed under a grant from the Natural Science and Engineering Research Council of Canada. Sections of this chapter were also published under the same title in *Fundamental Problems in Quantum Theory*, ed. Daniel M. Greenberger and Anton Zeilinger, Annals of the New York Academy of Science, vol. 755 (New York: New York Academy of Sciences, 1995) and in *Quantum Coherence and Decoherence—Foundations of Quantum Mechanics in Light of New Technology*, ed. K. Fujikawa and Y. A. Ono (Elsevier, 1996).

References

Aharonov, Y., P. G. Bergmann, and J. L. Lebowitz. 1964. *Phys. Rev.* 134B: B1410–B1416.

Aharonov, Y., et al. 1986. "Novel Properties of Preselected and Postselected Ensembles." In *New Techniques and Ideas in Quantum Measurement Theory*, vol. 480, Annals of the New York Academy of Sciences, ed. Daniel M. Greenberger. New York: New York Academy of Science. 417–21.

Aharonov, Y., J. Anandan, and L. Vaidman. 1993. *Phys. Rev.* A47: 4616–26.

Landauer, R., and T. Martin. 1994. *Reviews of Modern Physics* 66: 217–28.
Unruh, W. G. 1986. "Quantum Measurements." In *New Techniques and Ideas in Quantum Measurement Theory*, vol. 480, Annals of the New York Academy of Sciences, ed. Daniel M. Greenberger. New York: New York Academy of Science. 242–49.
———. 1994. *Phys. Rev.* A50: 882–87.
Von Neumann, J. 1955. *The Mathematical Foundation of Quantum Mechanics*. Trans. R. T. Beyer. Princeton: Princeton University Press.

Contributors

Guido Bacciagaluppi studied mathematics at the Swiss Federal School of Technology and the philosophy of science at Cambridge University. He is currently a British Academy postdoctoral fellow at Balliol College, Oxford, and faculty member in the department of philosophy. His main area of research is the philosophy of quantum mechanics.

Jeffrey Bub is professor of philosophy at the University of Maryland, College Park. He has published widely on the measurement problem and other interpretative issues of quantum mechanics. He is the author of *The Interpretation of Quantum Mechanics* (1974) and *Interpreting the Quantum World* (1997). His current area of research is quantum cryptography.

Rob Clifton is associate professor of philosophy at the University of Pittsburgh. He recently edited a volume entitled *Perspectives on Quantum Reality* and is currently coauthoring a book with Michael Dickson on modal interpretations of quantum mechanics.

Michael Dickson is assistant professor of history and philosophy of science at Indiana University. His research interests are in the philosophy of quantum mechanics. He has recently published the book *Quantum Chance and Nonlocality*.

Dennis Dieks studied theoretical physics at the University of Amsterdam. He obtained his Ph.D. at Utrecht University, with a discussion on the foundation of physics. He is currently professor of the foundations and philosophy of physics at Utrecht University.

Andrew Elby completed his doctoral work on the philosophy of physics at the University of California, Berkeley. He currently teaches at the high school and college levels, conducts education research, and occasionally ponders decoherence and modal interpretations.

Richard A. Healey is professor of philosophy at the University of Arizona. He previously taught at the University of California, Davis and Los Angeles. He has contributed many papers on the philosophy of physics to a variety of journals. His book, *The Philosophy of Quantum Mechanics: An Interactive Interpretation*, was published in 1989.

Geoffrey Hellman is professor of philosophy at the University of Minnesota and a member of the Center for the Philosophy of Science. He has published numerous articles on the philosophy of quantum mechanics and has contributed extensively to the philosophy of mathematics, including *Mathematics without Numbers: Towards a Modal-Structural Interpretation* (1989). He has held several scholars' awards from the National Science Foundation and has served on the editorial boards of leading scholarly journals, including *Philosophy of Science* and *Philosophia Mathematica.*

Meir Hemmo holds an Alon fellowship in the philosophy department, Haifa University. His doctoral thesis, concerning the foundations of non-collapsing quantum mechanics, was written at Cambridge University. He has contributed several papers to various journals and books. His current research focuses on questions related to the measurement problem, the role of decoherence, and the interpretation of probability.

Anthony J. Leggett is John D. and Catherine T. MacArthur Professor at the University of Illinois at Urbana-Champaign. His principal research interests are in the areas of condensed-matter physics, particularly high-temperature superconductivity, and the foundations of quantum mechanics.

Bradley Monton is a graduate student in philosophy at Princeton University.

Abner Shimony is professor emeritus of philosophy and physics at Boston University. Scientific inference, naturalistic epistemology, and the philosophy of physics have been his areas of specialization within philosophy, and the foundations of quantum mechanics have been his focus within physics. His aim has been to relate the two disciplines in a program he likes to call "experimental metaphysics." Many of his scientific and philosophical essays are collected in the two volumes entitled *Search for a Naturalistic World View* (1993). He served as president of the Philosophy of Science Association from 1995 to 1996.

William G. Unruh is a world-renowned physicist working in gravity and quantum theory. He is the founding director of the cosmology and gravity program at the Canadian Institute for Advanced Research. Currently, he is professor of physics at the University of British Columbia.

Pieter E. Vermaas studied theoretical physics and philosophy at the University of Amsterdam. The main topic of his Ph.D. research at Utrecht University concerned modal interpretations of quantum mechanics. Presently he is working on philosophy of technology at the Delft University of Technology.

Index